# Boston Studies in the Philosophy and History of Science

Volume 321

More information about this series at http://www.springer.com/series/5710

Jeroen van Dongen • Herman Paul
Editors

# Epistemic Virtues in the Sciences and the Humanities

Springer

*Editors*
Jeroen van Dongen
ITFA, Faculty of Science
University of Amsterdam
Amsterdam, The Netherlands

Herman Paul
Institute for History
Leiden University
Leiden, The Netherlands

ISSN 0068-0346 ISSN 2214-7942 (electronic)
Boston Studies in the Philosophy and History of Science
ISBN 978-3-319-48892-9 ISBN 978-3-319-48893-6 (eBook)
https://doi.org/10.1007/978-3-319-48893-6

Library of Congress Control Number: 2017953946



Printed on acid-free paper

This Springer imprint is published by Springer Nature
The registered company is Springer International Publishing AG
The registered company address is: Gewerbestrasse 11, 6330 Cham, Switzerland

# Contents

# Introduction: Epistemic Virtues in the Sciences and the Humanities

**Jeroen van Dongen and Herman Paul**

**Abstract** Epistemic virtues offer a promising angle for studying interaction between fields of research conventionally classified under the "sciences" and the "humanities." Given that virtues like objectivity, honesty, and accuracy are not confined to specific disciplines, they allow for comparative historical research between scientific fields as well as for histories of transfer, borrowing, and adaptation between disciplines. Such research, however, requires ample attention to what scientists in specific settings understood epistemic virtues to mean. Although the adjective refers to their role in facilitating the pursuit of epistemic aims (knowledge and understanding of reality), epistemic virtues are often imbued with moral, social, religious, and/or political meaning. If virtues specify the character traits marking a "scientific self," then scientific selfhood is never exclusively defined in epistemic terms.

**Keywords** Epistemic virtues • Epistemic vices • Virtue epistemology • Historical epistemology • History of knowledge • History of science • History of humanities • Scientific selfhood • Pierre Duhem

A few years ago, *Studies in History and Philosophy of Science* published a series of articles on the question whether Pierre Duhem (1861–1916), the French physicist, historian, and philosopher of science, could be considered a virtue epistemologist avant la lettre (Stump 2007; Ivanova 2010; Kidd 2011). Although virtue epistemology is a recent philosophical movement, committed to reviving the Aristotelian insight that obtaining knowledge requires exercise of intellectual virtues such as honesty, impartiality, and fairness (Zagzebski 1996; Baehr 2011), the authors observed that Duhem as early as the early twentieth century recognized a need for scientists to possess such qualities as rectitude, probity, impartiality, and "detachment from all interest and all passions" (Duhem 1991 [1915]: 43). Scientific work,

J. van Dongen (✉)
ITFA, Faculty of Science, University of Amsterdam, Amsterdam, The Netherlands

H. Paul
Institute for History, Leiden University, Leiden, The Netherlands

 
J. van Dongen, H. Paul (eds.), *Epistemic Virtues in the Sciences and the Humanities*,
Boston Studies in the Philosophy and History of Science 321,
https://doi.org/10.1007/978-3-319-48893-6_1

for Duhem, made a demand on the self: it required cultivation of qualities beneficial to acquiring knowledge and evaluating competing scientific theories. In Duhem's own words: "[I]t is not enough to be a good mathematician and skillful experimenter; one must also be an impartial and faithful judge" (Duhem 1954 [1914]: 218). Does this make Duhem a forerunner of virtue epistemology?

If Duhem associated the life of the mind with the exercise of intellectual virtues, so did many of his contemporaries in and outside of France. Drawing on ancient repertoires and consistent with a centuries-old tradition (Heyd 1995; Murphy and Traninger 2014; Demeter, Murphy and Zittel 2015), scholars encouraged their students to develop scientific character traits (Tannoch-Bland 1997; Eskildsen 2015; Manteufel 2016); subjected them to often ascetic technologies of the self (Levine 2002; Warwick 2003; Eskildsen 2016); cultivated "criticism," "objectivity," or "trustworthiness" as markers of scientific selfhood (Daston and Galison 2007; White 2009; Tollebeek 2011; Eskildsen 2013); inscribed themselves in narratives of virtue conquering vice (Paul 2012; Daston 2014; Engberts 2016); ascribed exemplary virtuousness to scientific role models (Bellon 2014; Schnicke 2015; Tollebeek 2015); and invoked standards of virtue and vice in book reviews and learned controversies (Paul 2016; Saarloos 2016). Even nineteenth-century language of methods, staunchly criticized by Duhem (1991 [1915], 44-47), not seldom appealed to character virtues beneficial to scientific research (Paul 2011, 2014). So, if Duhem anticipated virtue epistemology, cannot the same be said about many of his nineteenth- and early-twentieth-century colleagues?

Epistemology was, of course, not a field to which most scientists intended to contribute. When they invoked categories of virtue and vice, they mostly did so in order to articulate performance standards, scientific personae, or ideals of scientific selfhood. They used language of virtue and vice when appealing to norms of scientific conduct that had to be taught, practiced, and assessed. Until well into the twentieth century, language of virtue and vice served as a primary idiom for reflection on the scientific vocation and, especially, its implications for the scientific self (what did it take to be a scientist in terms of qualities that had to be cultivated or suppressed?). Only in more recent decades, language of skills and competences has come to supplement or even to replace this idiom of virtues and vices as a dominant vocabulary for addressing the personal qualities or dispositions needed for successful accomplishment of scientific work (Shapin 2008).

Yet if language of virtue referred to performance standards and idealized images of scholarly excellence, then to what extent can these virtues be classified as epistemic? How appropriate is the label of "epistemic virtues" for what virtue language meant to Duhem and his contemporaries? Much depends, of course, on what the adjective is supposed to mean. Virtue epistemologists and historians of science who have begun applying the category of epistemic virtues generally agree that epistemic virtues are virtues tout court. Epistemic virtues are "literal, not just metaphorical virtues," write Lorraine Daston and Peter Galison, to the extent that they require cultivation of character traits deemed necessary for scientific work. "Epistemic virtues earn their right to be called virtues by molding the self" (Daston and Galison 2007: 39, 41). Given that such work on the self has implications that transcend the quest for knowledge and understanding of reality, epistemic virtues always have more than only epistemic import. It would be wrong, therefore, to assume that epistemic virtues can be separated neatly from, say, moral virtues. At the same time, the adjective indicates the teleological orientation of the virtues in question: epistemic virtues are labeled epistemic because of their perceived relevance to the pursuit of epistemic aims. As Daston and Galison put it: "Epistemic virtues in science are preached and practiced in order to know the world, not the self" (2007: 40). So, despite the fact that epistemic virtues may manifest themselves in many domains at the same time, their principal intended locus of work, justifying the adjective, is the epistemic one (Creyghton et al. 2016).

Consistent with this relative priority of the epistemic, historians of science have studied epistemic virtues with an eye primarily on their aspirations in that realm. Although Daston and Galison note in passing that epistemic virtues are not seldom "parasitic upon religious impulses to discipline and sacrifice" (2007: 41), their book de facto keeps silent about religious subtexts: it focuses exclusively on different ways in which objectivity was believed to serve epistemic aims. Likewise, when Martin Mulsow examines how the Enlightenment ideal of sapere aude ("dare to know") was translated into epistemic virtues such as intellectual courage, he observes that this program had a specific anti-theological agenda. Yet his focus remains on standards of knowledge propagated by eighteenth-century Enlightenment authors, not on religious ideals or social visions implied in their calls for independent thinking (Mulsow 2012: 220-233).

Duhem is a case in point, too. Shortly before his death, in the embittered context of World War I, he published *La Science Allemande* (1915), in which he contrasted a Pascalian *esprit de finesse*, typical for French science, with German "arrogance," "presumptuousness," and "inordinateness" (Duhem 1991 [1915]: 52). *La Science Allemande* was a defense of French "intuition" against a "mathematical mind" characterized by industriousness, meticulousness, discipline, and submission (Duhem 1991 [1915]: 116-118). In this wartime attempt at "national defense" (Duhem 1991 [1915]: 6), then, epistemic virtues and vices took on political color. Yet, in the secondary literature on Duhem, the question how his scientific nationalism imbued epistemic virtues with political meaning ("German thoroughness," "French lucidity") has not remotely been studied as closely as his reflections on "good sense" (*bon sens*) as "a guide to theory choice" (Ivanova 2010) – even though Duhem's Catholicism has received ample attention (Jaki 1991; Martin 1991; Kragh 2008).

As long as historians of science restrict their subject matter to knowledge production in the past – what did scholars do in order to increase their knowledge of reality? – this epistemic focus hardly needs defense. The situation changes, however, as soon as historians of science take a "cultural turn," as many in the past few decades have done (Clark 2015). Studying science as a cultural activity amounts to treating it, in Steven Shapin's celebrated phrase, "as if it was produced by people with bodies, situated in time, space, culture, and society, and struggling for credibility and authority" – and hence as "never pure" (Shapin 2010). Although cultural historians of science do not cease to be interested in the epistemic aspirations of scientists past and present, they expand the research agenda by raising the question how such epistemic goals interacted with other aims. They examine, for example, how epistemic inquiry was justified and conditioned in political, religious, or aesthetic terms, what sort of contexts it was supposed to require, and what kind of meaning science had for different audiences.

Although such a cultural history of epistemic virtues is only beginning to emerge, there is no lack of examples suggesting its relevance. Objectivity, to mention Daston's and Galison's example once again, rose to great heights in the decades around 1900. Yet, this did not imply that the virtue was accessible to everyone. Late-nineteenth-century German historians, most of whom identified as Protestants, regularly excluded Jews and Roman-Catholics from full membership of the profession by arguing that objectivity was a fruit of Luther's Reformation, which they interpreted in liberal vein as a protest against coercion of conscience (Krill 1962). This single example alone already illustrates that a history of objectivity intent on reconstructing the meanings associated with this particular virtue has to take into account not only changing scientific practices but also the history of religion (liberal Protestantism in particular), the German political situation shortly after the Kulturkampf, and the long history of academic anti-Semitism.

A related observation is that social, political, economic, or religious contexts could shape what counted as a virtue or vice. When mid-nineteenth-century American chemists offered scientific advice to oil and coal companies, some critics condemned this as a surrender to Mammon – a vice par excellence (Lucier 2008). By contrast, in the more advanced capitalist context of the late twentieth century, an ability to engage in productive cooperation with industrial partners was considered most virtuous and an evident service to society (Shapin 2008). Likewise, while patristic and medieval theologians had unanimously condemned curiosity as a vice, closely related to pride and self-aggrandizing, in the course of the seventeenth century, through the influence of Francis Bacon and Thomas Hobbes, among others, curiosity gradually acquired the status of epistemic virtue (Daston 1995b; Harrison 2001; Benedict 2001).

So, although epistemic virtues were supposed to serve epistemic aims, the range of meanings that scientists associated with them reached widely beyond the epistemic sphere. No matter how much virtues seemed finely attuned to epistemic aims, at closer inspection they not seldom turned out to mix closely with political agendas, rely on gendered notions of masculinity, or interfere with moral and aesthetic values (e.g., Benthien and Martus 2006; Murphy and Traninger 2014; Schnicke 2015). A history of epistemic virtues narrowly focused on their epistemic aspirations therefore leaves out much of their complexity. From a cultural historical point of view, the scope needs to be broadened so as to acknowledge multiple, overlapping, and conflicting layers of meanings. Epistemic virtues, in other words, are better studied fully placed within the broader cultural realms they are necessarily part of – that is, in cultural backgrounds invoked by historians to explain the patterns of meaning historically attributed to their objects of study. The first and foremost valuable insight that historians of science and the humanities can offer virtue epistemologists who have discovered Duhem as their predecessor is a such a "local" or "cultural" (Burke 2004: 3) reading of the virtues that Duhem and his contemporaries deemed necessary for scientific work.

A major additional benefit of such a cultural history of epistemic virtues is that it may also contribute to a more interdisciplinary or even "post-disciplinary" history of knowledge – one that transcends traditional disciplinary boundaries and the local (Bod et al. forthcoming; Paul forthcoming). The history of science, especially in the English-speaking world, has usually been focused on the natural sciences. Recently, calls for more comparative and integrated study of the "three cultures" – natural sciences, social sciences, and humanities – are heard (Bod 2015; Daston and Most 2015). Epistemic virtues are a promising topic and historiographic tool for such studies, because they were often shared, transferred, traded, and borrowed across disciplinary boundaries (e.g., Paul 2012; Turner 2014, Tai and van Dongen 2016). Also, perhaps especially in the social sciences and humanities, scientific selfhood was frequently defined in relation to scientific "others" that served as models to be imitated or avoided (Lepenies 1985; Backhouse and Fontaine 2010). Unlike discipline-specific research practices, then, epistemic virtues, understood as markers of scientific selfhood, lend themselves particularly well to comparison across the academic spectrum – especially if the cultural contexts that gave meaning to those virtues are taken into account.

Obviously, this volume does not offer a full history of scientific selfhood, but it contributes to such a history by drawing attention to qualities that scientists were supposed to embody and to the rich layers of meanings often associated with those qualities. In particular, this volume seeks to contribute to a cultural history of scientific selfhood by crossing disciplinary divides itself. It examines epistemic virtues in a variety of political, cultural, and academic settings. Drawing on case studies from fields as diverse as physics, chemistry, history, philology, and Oriental studies, the chapters in this volume discuss why some epistemic virtues were more widespread than others; how virtues related to vices and sociopolitical contexts helped shape catalogs of virtues and vices; on what grounds scientists justified their epistemic virtues; how such virtues interacted with moral, aesthetic, and religious values; how they influenced the knowledge produced; and why virtues as well as scientists who were perceived as embodying them were more contested in some fields than in others.

In his opening contribution, Ian James Kidd highlights the metaphysical subtexts of such epistemic virtues as humility. If epistemic confidence in science changes, so do conceptions of what constitutes humble scientific conduct. This is particularly evident from the nineteenth-century Victorian scholars Kidd zooms in on. Highly confident scientific naturalists such as John Tyndall illustrate that metaphysical commitments shape what the virtue of humility looks like and, consequently, what counts as humble or hubristic scientific conduct.

As Jessica Wang and Matthew Stanley show in their subsequent contributions, much the same is true for religious, moral, and aesthetic commitments. When the Cold War physicists Merle Tuve and Robert Wilson identified with a pure science ideal, they did so out of deep anxieties about their personal role in the increasingly large-scale and instrumental big science enterprise that American science was becoming. Wang then demonstrates that moral conviction and personal religious belief were closely tied to epistemic virtues. Yet this does not imply that there is a one-to-one relation between epistemic and religious virtues. As Stanley shows for the virtue of "persistence" as advocated by James Clerk Maxwell, a nineteenth-century conservative, and Arthur Eddington, an early-twentieth-century liberal Quaker, one and the same virtue could be defended from very different religious perspectives.

In the case of Albert Einstein, too, religious sentiment was never far away. In his efforts to influence the direction of the emerging subdiscipline of theoretical physics, Einstein made deliberate use of virtue language, not least when presenting himself as a virtuous scientist. As Jeroen van Dongen shows, Einstein even couched his position in evocative Spinozist accounts of how attempts to "unify" the natural forces would be connected to personal deliverance, thereby persistently conflating the epistemic and the moral.

Similar conflations can be observed in how nineteenth-century European Orientalists evaluated the scholarly achievements of Heinrich Ewald and Reinhart Dozy. Christiaan Engberts and Herman Paul argue that the "dogmatism" and "rashness" associated with these German and Dutch scholars were epistemic vices as much as they were social and religious ones: dispositions perceived as detrimental to scholarly work were indeed often hard to distinguish from social, moral, or religious vices.

While Ewald and Dozy served as epitomes of vice, the Göttingen professors Wilhelm Weber, Friedrich Wöhler, and Georg Waitz were held in high regard. But as Herman Paul shows in a comparative analysis of their posthumous reputations, language of virtue was more prominent in some disciplines (history) than in others (physics, chemistry). Whether or not scholars felt a need to present their deceased teachers as paragons of epistemic virtue depended, among other things, on specific disciplinary contexts. The more a discipline was divided over the question what it took to be a good scholar in terms of habits, skills, and virtues, the more important was language of virtue and vice in evaluating the relative merits of high-profile academics.

Disagreement over epistemic virtues could also exist within disciplines, as Léjon Saarloos shows in his chapter. Focusing on John Tyndall's debates with Peter Guthrie Tait on the use of the "imagination" in science, Saarloos shows that Tait urged restraint, whereas Tyndall emphasized intellectual courage – an epistemic virtue insufficiently recognized in scholarship that has equated Tait-style asceticism with epistemic virtuousness tout court. Raf de Bont, in turn, makes clear that different loci, even within a single, small discipline, could require rather different virtues. He shows that scientific nature preservationists needed more than the travel and hunting skills required for fieldwork in Africa or Asia: besides the "field," there was the "office" – and these epistemic sites required different catalogs of epistemic virtues. Additionally, de Bont's chapter offers further corroboration of the view that virtues conducive to epistemic aims are often impossible to separate from virtues serving moral aims, such as the goal of nature preservation.

This raises the question whether and how virtues can serve as connections between seemingly different fields of activity. In his chapter on the Russian-American linguist Roman Jakobson, Bart Karstens shows not only that Jakobson's structuralism was influenced by formalist poetry but also, more generally, that virtues connected his views on language, science, and art. Specifically, as Karstens puts it, "studying Jakobson through the prism of the pursuit of epistemic virtues helps explain how he could draw connections between a wide variety of fields, including musicology, physics, genetics, communication theory, vivisection, and chemistry."[1] In his final chapter, Ad Maas suggests that connections existed between academic research and the social-political order of the nineteenth-century Netherlands: the latter appeared to condition what came to be regarded as important epistemic virtues in the former. Zeroing in on the liberal reforms enacted in 1848 by the philosopher-cum-historian Johan Rudolph Thorbecke, Maas discusses how different conceptions of the state resonated with different understandings of how historians should conduct their profession.

What can be learned, then, from this volume? When crossing disciplinary boundaries in the historiography of scholarship, epistemic virtues appear to be immensely useful tools. On the one hand, they inform and direct scientific practice. As Peter Galison argues, "epistemic virtues are part of the constitutive core of science. There is nothing secondary about mathematical rigor … or objectivity" (Harman and Galison 2008: 563). Indeed, what makes epistemic virtues worth studying is that they negotiate ideals and practices. Given that they also interact and travel across disciplinary boundaries, they offer a promising angle for studies of knowledge transfer and for what we called a "post-disciplinary" history of the sciences and the humanities, particularly when such historiography is conducted at a level beyond the individual case study – a level of engagement that is greatly sought after in today's history of science (Daston and Galison 2008). Finally, as many chapters in this volume show, epistemic virtues help situate the scientific self in its cultural contexts. Time and again, scientists define their vocation in terms with religious, moral, political, and/or metaphysical overtones, thereby illustrating that epistemic virtues can never be fully separated from religious, moral, or political ones (Creyghton et al. 2016). The history of scientific selfhood to which this volume intends to contribute is one that is sensitive to precisely such culturally specific meanings associated with epistemic virtues and vices.

**Acknowledgments** The papers collected in this volume are a selection of presentations given at the conference "Epistemic Virtues in the Sciences and the Humanities," held on 27–28 August 2015 at the Descartes Centre for the History and Philosophy of the Sciences and the Humanities at Utrecht University. The editors would like to thank the Descartes Centre for its most generous support of this event and are very grateful to Descartes' Ariane den Daas for her assistance in organizing the event. Also, they thankfully acknowledge financial support from the Netherlands Organisation for Scientific Research (NWO).

[1] In a similar vein, Tai and van Dongen (2016) use Anton Pannekoek's epistemic virtues to connect his Marxism and astronomy – two fields of activity that Pannekoek himself always argued to be entirely separate.

## Bibliography

Backhouse, Roger E. and Philippe Fontaine, eds. 2010. *The History of the Social Sciences since 1945*. Cambridge: Cambridge University Press.
Baehr, Jason. 2011. *The Inquiring Mind: On Intellectual Virtues and Virtue Epistemology*. Oxford: Oxford University Press.
Bellon, Richard. 2014. There is Grandeur in this View of Newton: Charles Darwin, Isaac Newton and Victorian Conceptions of Scientific Virtue. *Endeavour* 38, 222–234.
Benedict, Barbara M. 2001. *Curiosity: A Cultural History of Early Modern Inquiry*. Chicago: University of Chicago Press.
Benthien, Claudia and Steffen Martus, eds. 2006. *Die Kunst der Aufrichtigkeit im 17. Jahrhundert.* Tübingen: Max Niemeyer.
Bod, Rens. 2015. A Comparative Framework for Studying the Histories of the Humanities and Science. *Isis* 106, 367–377.
Bod, Rens, Jaap Maat, and Thijs Weststeijn, eds. 2012. *The Making of the Humanities*, vol. 2. Amsterdam: Amsterdam University Press.
_____. 2014. *The Making of the Humanities*, vol. 3. Amsterdam: Amsterdam University Press.
Bod, Rens, Jeroen van Dongen, Sjang ten Hagen, Bart Karstens, and Emma Mojet. Forthcoming. *The Flow of Cognitive Goods: Towards a Post-Disciplinary Perspective on the History of Knowledge.*
Burke, Peter. 2004. *What is Cultural History?* Cambridge: Polity Press.
Clark, John F. M. 2015. Intellectual History and the History of Science. In *A Companion to Intellectual History*, ed. Richard Whatmore and Brian Young. Chichester: Wiley Blackwell, 155–169.
Creyghton, Camille, Pieter Huistra, Sarah Keymeulen, and Herman Paul. 2016. Virtue Language in Historical Scholarship: The Cases of Georg Waitz, Gabriel Monod and Henri Pirenne. *History of European Ideas* 42, 924–936.
Daston, Lorraine. 1995a. The Moral Economy of Science. *Osiris* 10, 2–24.
_____. 1995b. Curiosity in Early Modern Science. *Word and Image* 11, 391–404.
_____. 2014. Objectivity and Impartiality: Epistemic Virtues in the Humanities. In Bod, Maat, and Weststeijn 2014, 27–41.
Daston, Lorraine and Peter Galison. 2007. *Objectivity*. New York: Zone Books.
_____. 2008. Objectivity and Its Critics. *Victorian Studies* 50, 666–677.
Daston, Lorraine and Glenn W. Most. 2015. History of Science and History of Philologies. *Isis* 106, 378–390.
Demeter, Tamás, Kathryn Murphy, and Claus Zittel, eds., 2015. *Conflicting Values of Inquiry: Ideologies of Epistemology in Early Modern Europe*. Leiden: Brill.
Duhem, Pierre. 1954 [1914]. *The Aim and Structure of Physical Theory*, trans. Philip P. Wiener. Princeton: Princeton University Press.
_____.1991 [1915]. *German Science: Some Reflections on German Science: German Science and German Virtues*, trans. John Lyon. La Salle, IL: Open Court.
Engberts, Christiaan. 2016. Gossiping about the Buddha of Göttingen: Heinrich Ewald as an Unscholarly Persona. *History of Humanities* 1, 371–385.
Eskildsen, Kasper Risbjerg. 2013. Inventing the Archive: Testimony and Virtue in Modern Historiography. *History of the Human Sciences* 26, no. 4, 8–26.
_____. 2015. Private Übungen und verkörpertes Wissen: Zur Unterrichtspraxis der Geschichtswissenschaft im neunzehnten Jahrhundert. In Martin Kintzinger and Sita Steckel, eds., *Akademische Wissenskulturen: Praktiken des Lehrens und Forschens vom Mittelalter bis zur Moderne.* Bern: Schwabe, 143–161.
_____. 2016. Scholarship as a Way of Life: Character and Virtue in the Age of Big Humanities. *History of Humanities* 1, 387–397.

Harman, Oren and Peter L. Galison. 2008. Epistemic Virtues and Leibnizian Dreams: On the Shifting Boundaries between Science, Humanities and Faith. *The European Legacy* 13, 551–575.

Harrison, Peter. 2001. Curiosity, Forbidden Knowledge, and the Reformation of Natural Philosophy in Early Modern England. *Isis* 92, 265–290.

Heyd, Michael. 1995. *Be Sober and Reasonable: The Critique of Enthusiasm in the Seventeenth and Early Eighteenth Centuries*. Leiden: Brill.

Ivanova, Milena. 2010. Pierre Duhem's Good Sense as a Guide to Theory Choice. *Studies in History and Philosophy of Science* 41, 58–64.

Jaki, Stanley L. 1991. *Scientist and Catholic: Pierre Duhem*. Front Royal, VA: Christendom Press.

Kidd, Ian James. 2011. Pierre Duhem's Epistemic Aims and the Intellectual Virtue of Humility: A Reply to Ivanova. *Studies in History and Philosophy of Science* 42, 185–189.

_____. 2014. Was Sir William Crookes Epistemically Virtuous? *Studies in History and Philosophy of Biological and Biomedical Sciences* 48, 67–74.

Kragh, Helge. 2008. Pierre Duhem, Entropy, and Christian Faith. *Physics in Perspective* 10, 379–395.

Krill, Hans-Heinz. 1962. *Die Rankerenaissance: Max Lenz und Erich Marcks: Ein Beitrag zum historisch-politischen Denken in Deutschland, 1880–1935*. Berlin: Walter de Gruyter.

Lepenies, Wolf. 1985. *Die drei Kulturen: Soziologie zwischen Literatur und Wissenschaft*. Munich: Hanser.

Lucier, Paul. 2008. *Scientists and Swindlers: Consulting on Coal and Oil in America, 1820–1890*. Baltimore, MD: Johns Hopkins University Press.

Manteufel, Katharina. 2016. A Three-Story House: Adolf von Harnack and Practices of Academic Mentoring around 1900, *History of Humanities* 1, 355–370.

Martin, R. N. D. 1991. *Pierre Duhem: Philosophy and History in the Work of a Believing Physicist*. La Salle, IL: Open Court.

McKenna, Mark. 2011. *An Eye for Eternity: The Life of Manning Clark*. Carlton: Melbourne University Publishing.

Mulsow, Martin. 2012. *Prekäres Wissen: Eine andere Ideengeschichte der Frühen Neuzeit*. Berlin: Suhrkamp.

Murphy, Kathryn and Anita Traninger, eds. 2014. *The Emergence of Impartiality*. Leiden: Brill.

Paul, Herman. 2011. Distance and Self-Distanciation: Intellectual Virtue and Historical Method around 1900. *History and Theory* 50, no. 4, 104–116.

_____. 2012. The Scholarly Self: Ideals of Intellectual Virtue in Nineteenth-Century Leiden. In Bod, Maat, and Weststeijn 2012, 397–411.

_____. 2014. Manuals on Historical Method: A Genre of Polemical Reflection on the Aims of Science. In Bod, Maat, and Weststeijn 2014, 171–182.

_____. 2016. The Virtues and Vices of Albert Naudé: Toward a History of Scholarly Personae. *History of Humanities* 1, 327–338.

_____. Forthcoming. *German Thoroughness in Baltimore: Epistemic Virtues and National Stereotypes*.

Philips, Denise. 2015. Francis Bacon and the Germans: Stories from when "Science" Meant "Wissenschaft." *History of Science* 53, 378–394.

Saarloos, Léjon. 2016. Virtue and Vice in Academic Memory: Lord Acton and Charles Oman, *History of Humanities* 1, 339–354.

Schnicke, Falko. 2015. Rituale der Verkörperung: Seminarfeste und Jubiläen der Geschichtswissenschaft des 19. Jahrhunderts. *Zeitschrift für Geschichtswissenschaft* 63, 337–358.

Shapin, Steven. 2008. *The Scientific Life: A Moral History of a Late Modern Vocation*. Chicago: University of Chicago Press.

_____. 2010. *Never Pure: Historical Studies of Science as if It Was Produced by People with Bodies, Situated in Time, Space, Culture, and Society, and Struggling for Credibility and Authority*. Baltimore, MD: Johns Hopkins University Press.

Stump, David J. 2007. Pierre Duhem's Virtue Epistemology. *Studies in History and Philosophy of Science* 38, 149–159.
Tai, Chaokang and Jeroen van Dongen. 2016. Anton Pannekoek's Epistemic Virtues in Astronomy and Socialism: Personae and the Practice of Science. *Low Countries Historical Review* 131, 55–70.
Tannoch-Bland, Jennifer. 1997. Dugald Stewart on Intellectual Character. *The British Journal for the History of Science* 30, 307–320.
Tollebeek, Jo. 2011. *Men of Character: The Emergence of the Modern Humanities*. Wassenaar: Netherlands Institute for Advanced Study in the Humanities and Social Sciences.
_____. 2015. Commemorative Practices in the Humanities around 1900, *Advances in Historical Studies* 4, 216–231.
Turner, James. 2014. *Philology: The Forgotten Origins of the Modern Humanities*. Princeton: Princeton University Press.
Warwick, Andrew. 2003. *Masters of Theory: Cambridge and the Rise of Mathematical Physics*. Chicago: University of Chicago Press.
White, Paul. 2009. Darwin's Emotions: The Scientific Self and the Sentiment of Objectivity. *Isis* 100, 811–826.
Zagzebski, Linda Trinkaus. 1996. *Virtues of the Mind: An Inquiry into the Nature of Virtue and the Ethical Foundations of Knowledge*. Cambridge: Cambridge University Press.

# Confidence, Humility, and Hubris in Victorian Scientific Naturalism

**Ian James Kidd**

**Abstract** Most historians explain changes in conceptions of the epistemic virtues and vices in terms of social and historical developments. I argue that such approaches, valuable as they are, neglect the fact that certain changes also reflect changes in metaphysical sensibilities. Certain epistemic virtues and vices are defined relative to an estimate of our epistemic situation that is, in turn, defined by a broader vision or picture of the nature of reality. I defend this claim by charting changing conceptions of the virtue of epistemic humility in nineteenth century intellectual culture, focusing on claims about the humility or hubris of the Victorian scientific naturalists.

**Keywords** Epistemic virtues • Epistemic vices • Virtue epistemology • Epistemic humility • Epistemic confidence • Epistemic hubris • Ludwig Wittgenstein • Herbert Spencer • Auguste Comte • Charles Sanders Peirce

## Introduction

It is clear that the epistemic virtues and vices recognised by scientists and humanists change over time. Some virtues come and go, even as others remain fairly stable, and others still can change status, being an excellence of epistemic character at one time, only to later become a vice. A good example is curiosity, barely mentioned by the ancient Greeks, castigated as a vice during much of the medieval period, but gradually rehabilitated, from the early modern period onwards, as one of the most crucial 'virtues of the mind' (see Zagzebski 1996). Similar stories about the changing status of other epistemic character traits can been told, even if those histories tend to be neglected by virtue epistemologists. Happily, more studies are emerging that explore the historical development of various virtues in social, religious, and intellectual context (see Cooper 2002 and Pardue 2013).

But the majority of the existing studies of changes in the sorts and types of epistemic virtues and vices have tended to focus on social and historical factors.

I.J. Kidd (✉)
University of Nottingham, Nottingham, UK
e-mail: Ian.Kidd@nottingham.ac.uk



J. van Dongen, H. Paul (eds.), *Epistemic Virtues in the Sciences and the Humanities*, Boston Studies in the Philosophy and History of Science 321,
https://doi.org/10.1007/978-3-319-48893-6_2

The shifting conceptions of the 'scientific self' depicted by Steven Shapin (2008) are characterised by the virtues appropriate to culturally mandated values. The Godly Naturalist of the eighteenth century is pious and humble, while the late modern Venture Capitalist is confident, ambitious. Similarly, Lorraine Daston and Peter Galison (2007) chart changing 'regulative visions' of the virtues of objective enquirers. A 'mechanically objective' scientist is disciplined, restrained, while another, embracing 'trained judgment', is discerning, creative. In each case, changes in tables of the virtues are explained in relation to social and historical developments – new technologies, representational techniques, economic imperatives, intellectual developments, and so on.

Such studies are valuable, but neglect a further type of factor crucial to explanations of historical shifts in the epistemic virtues and vices. I refer to changes in the *metaphysical sensibilities* of a culture, to guiding visions of the nature of reality. One might expect interest in the metaphysics of epistemic virtue from those philosophers interested in epistemic virtues – virtue epistemologists. But, surprisingly, very few of them explicitly explore the grounding of the virtues of the mind in such things as worldviews, or indeed to their history. Honourable exceptions are Bob Roberts and Jay Wood, who throughout their book, *Intellectual Virtues*, remark that '(epistemic) virtue and vice concepts' are tacitly 'indexed' to 'metaphysical commitments', 'world views', or a 'conception of human nature'. The intelligibility and salience of certain virtues and vices can only be explained by reference to the particular 'worldview' or 'metaphysical background', which those epistemic concepts 'presuppose' (2007, 155, 82, 22). When an account of a vice acknowledges its historical and metaphysical grounding, call it a *deep conception of epistemic vice* (see Kidd MS Unpublished).

The changing fortunes of curiosity, gestured to earlier, offer a good example of how a certain epistemic character trait requires explanatory appeal to shifting metaphysical visions. In Hans Blumenberg's account, many medievals followed Augustine in associating *curiositas* with spiritually objectionable attitudes and ambitions. Enthusiasm for enquiry into mundane matters was liable to fuel corrupting pleasure in our 'cognitive accomplishments', leading to various vices – 'cognitive arrogance', 'the danger of self-admiration', or a hubristic sense of 'autonomous cognitive security' (1983, 286, 310). Such vices are hardly compatible with the cultivation of a properly pious relationship with God, the primary good for human beings. If so, curiosity offers a case where the normative status of an epistemic character trait – as virtue or as vice – is determined relative to a metaphysical vision or worldview.

Without an appeal to a Christian worldview, argues Blumenberg, it is difficult to give a satisfying account of why *curiositas* came to be 'enrolled in the catalog of vices' (1983, 309). Indeed, it was no coincidence that curiosity's subsequent rehabilitation occurred at the same time as the emergence of new, rival visions of reality, joint products of humanism and natural philosophy. Such visions valorised human powers, encouraging potent new epistemic ambitions and thereby redeeming character traits, like curiosity, conducive to their pursuit.

My focus in this chapter is on a different virtue and a later stage in history. Using the remarks of Roberts and Wood as my starting point, I show that a full explanation of conflicting nineteenth century conceptions of the virtue of epistemic humility must invoke rival metaphysical visions – a claim I illustrate with the case of Victorian scientific naturalism.

## Epistemic Confidence

I start with an account of the virtue of epistemic humility, which I argue is a virtue that registers an appreciation of the conditional status of our epistemic confidence – the confidence invested in activities aimed at the acquisition, assessment, and application of knowledge and other epistemic goods (see further Kidd 2015, 2016a).

Central to this account is the idea that any given assertion, belief, or conviction will rest upon some set of *confidence conditions*. By that term, I mean the cognitive, practical, or other conditions that ought to be fulfilled if an agent is to have warranted confidence in some epistemic action or commitment. Perhaps confidence would require the agent to have certain sorts of specialist knowledge (*cognitive*) or an ability to perform certain actions (*practical*) or access to certain objects or places (*material*). A historian's confidence in some claim may be conditional upon their being trained in certain interpretive techniques, having access to a certain archive of documents, and so on. The specific types of confidence conditions will vary according to the specific actions and commitments and the contexts in which they are made – a fact reflected in the complexity of our social and epistemic life.

It is useful to distinguish three levels at which confidence conditions can operate. The first is *agential confidence*, those particular to the specific epistemic agent – perhaps in their general cognitive capacities, knowledge of a subject, or their 'self-trust' in their status as one able to pursue and attain truth. But agents must have confidence in others, too, so the second level is that of *collective confidence*, which is invested in both other agents and the structures that organise their activities – disciplines, communities, institutions, and so on. It might be a confidence in the people who teach and train us, or in the sense of fair play of one's critics, or in the robustness of shared practices, such as anonymous peer review or data-gathering.

As these examples make clear, collective confidence is complex. Colleagues can let us down, shared epistemic practices can be abused, and institutions can be corrupted. The virtue of epistemic humility therefore builds in, at the ground level, an acute sense of the fact that epistemic confidence is conditional, complex, contingent, and therefore *fragile*.

The philosophical literature tends to focus on the agential and collective aspects of our epistemic life. But there is a third, final level: that of *deep confidence*, the more tacit sort invested, by agents and collectives, in something broader and more abstract. Objects of deep confidence might include a project of enquiry – such as rational theology or natural science – or in the richness of a shared inheritance of thought and sensibility. Other objects take more 'theoretical' forms: perhaps a deep confidence in the fundamental intelligibility of reality, for instance, or the possibility of progress in a project of enquiry, without which enquiry would be futile and frustrating.

Such objects of deep confidence might seem too obvious to be worth specifying. Yet they are perennially disputed throughout our intellectual history and should not be taken for granted. Indeed, many of the deep rhythms of debate that run through that history can be interpreted as disputes about the sorts of deep confidence to which human beings can and should aspire.

Two points relevant to my later discussion are worth noting. First, the levels at which epistemic confidence operates – the agential, collective, and deep – are not sharply demarked from one another. On the contrary, they interpenetrate and interact, with changes at one level having ramifications for the others. A loss of deep confidence in philosophy, for instance, can precipitate a loss of confidence in its particular practitioners and communities. Early feminist histories of Western philosophy reflected a sense of alienation from a tradition perceived as a 'repository of misogynist ideas and ideals', recalls Genevieve Lloyd, later succeeded by projects of 'appropriation' that helped to restore confidence – for instance, in its capacity to 'enrich cultural self-understanding' (2000, 245). Second, the level of 'deep confidence' is indeed deep in the sense of 'fundamental', since the agential and collective levels sit on top of it. The fundamentality of deep confidence is central to my later discussion of the claims of nineteenth century scientific naturalists to be exemplars of epistemic humility.

With these remarks on confidence in place, I now give an account of humility.

## Epistemic Humility

I suggested that humility is a virtue marked by special sensitivity to the conditional status of the confidence that is essential to our epistemic life. But to merit the status of a virtue, taken as an excellence of character to be cultivated and admired, humility must do further work. It is only going to be virtuous to be epistemically humble if doing so makes a difference to our epistemic activities and, therefore, to our life.

The virtue of humility is structured around two pairs of components. The first is a disposition to recognise the fulfilment of the confidence conditions that are relevant to a given epistemic act or commitment—an assertion, say, or a conviction—in the ways described in the last section. A humble enquirer is disposed, that is, to recognise the sorts of conditions that might underlie or 'go with' that assertion or conviction, and whether or not they do or could fulfil them, either alone, as they stand, or with others, perhaps after further training. It is this sort of humility that is reflected in the idea of 'knowing (or acting within) one's limits'. But this is not, in itself, enough to make up a virtue: one could imagine a person who was perfectly good at *recognising* confidence conditions, but who consistently ignored them by 'holding forth' regardless – a tendency, surely, of an arrogant person.

So a second component is needed that connects the recognition of confidence conditions to actual changes in how one conducts oneself epistemically. This is a disposition to *regulate* one's epistemic conduct in the light of one's changing fulfilment of relevant confidence conditions. A humble enquirer is disposed, that is, to actively regulate their epistemic conduct, their personal ways of engaging in epistemic activities. Perhaps they take care to carefully qualify claims to reflect the degree of confidence they can justify, sharing Montaigne's fondness for words and phrases that 'soften and moderate' the typical 'rashness' of our speech – 'perhaps', 'I think', 'as far as I know' (1991, 1165). Perhaps they regulate their ambitions by ensuring the epistemic projects they commit to are ones whose constituent conditions they could fulfil – aspiring to *contribute* to, rather than *define*, an area of study.

Taken together, a humble enquirer is disposed to actively regulate their epistemic conduct by recognising and appropriately responding to the complex economy of confidence upon which their activities and projects rely. Since I have elsewhere given a fuller account of the virtue of epistemic humility (Kidd 2015, 2016a), I will here confine myself to two general remarks. The first is that it is quite consistent with a latent modern conception of humility, and the rhetorics and metaphors associated with it. Think of talk of 'getting a good grounding' in a new subject or an epistemological vocabulary of 'foundations'. Second, this account is domain-neutral, applying to scientific and humanistic enquiry, and ordinary and complex sorts of epistemic activity. A person can be epistemically humble, in the sense described, whether they are measuring solar flares, explaining the rise of fascism, or debating the merits of glam rock over drinks with friends. This neutrality is a deliberate feature of the account in its general form, but humility changes with time. One of the reasons why is that, over time, the objects or modes of deep confidence to which a humble enquirer is attuned change. These changes are, fundamentally, ones of 'worldview'.

I now say more about the relationship of deep confidence to metaphysical visions, using the case of the virtue of epistemic humility in the context of nineteenth century scientific naturalism.

## Deep Confidence

An enquirer with the virtue of epistemic humility will have sensitivity to the deep confidence conditions that are relevant to their epistemic conduct. It seems likely that much of their life will not require constant and intense focus on the deeper conditions: but, at times, it will and often for the reason that deep confidence can change, either being gained or, indeed, lost.

It is clear that deep confidence can and has taken many forms throughout intellectual history, too many to survey in any detail in this chapter, but certain general sorts stand out. To start with, there is a confidence in the power of human reason to ascertain truths of a suitably profound or complex sort about the world, of the sort classified with the capacious label of 'rationalism'. Next is a confidence in the methods and ambitions of rational theology or natural science, of a sort that can inspire people to pursue them, despite disappointments, obstacles, and failures. Another is confidence in the ultimate intelligibility of reality – of a trust that the world has an order, if not purpose, accessible by our epistemic capacities.

Several other general forms of deep confidence exist, but these suffice to illustrate two crucial points. One is that these diverse forms or modes of deep confidence can relate to and shape one another in complicated ways—for instance, deep confidence in science as a source of truths about reality (or significant aspects of it, anyway) has, historically, been closely related to a confidence in its progressiveness. This is recognized by that energetic group of critics – antirealists, constructivists, relativists – who have spent much of the last 40 years stridently challenging both components of this double mode of deep confidence in science.

The other crucial point is that deep confidence is typically contingent, in the sense of its being available only if certain conditions are fulfilled. The Biblical myth of the 'Fall of Man' that was later developed into postlapsarian theologies, argues the historian Peter Harrison, 'precipitated a crisis of confidence' in the cognitive powers of human beings and traditional sources of knowledge. If, epistemologically, 'error was … equated with sin', then we 'fallen' human beings inherited, along with original sin, a 'proneness to error', ignorance, and confusion. But responses to this 'crisis' varied, according to theological conviction. Augustine invoked divine 'grace' as a 'cure' for the 'inherited malady of sin', while Aquinas argued that our 'rational soul', created by God, contains its own 'light of natural reason', which cannot be 'forfeit', even after the fall (Harrison 2007, 4, 39, 43). Such appeals to divine grace or 'natural reason' can be understood as strategies for the fulfilment of deep confidence conditions, particular to the resources of medieval European Christian culture.

Early modern thinkers inherited similar worries about natural 'infirmities' and 'weaknesses' of the human mind. Their response was enthusiasm for 'regimens of the mind', directed to the 'cure and cultivation' of our corrupted yet restorable minds. Sorana Corneanu argues that 'regimens' of this sort were understood to be incumbent on a person, 'as an individual, as a member of a community, and as a creature in relation to its deity'. Such 'epistemic modesty' operates at agential, collective, and deep levels, and is ultimately rooted in an 'anthropological conception', dictating 'what the virtuous knower does and is like', itself embedded in a wider 'vision' of reality (Corneanu 2011, 2, 99, 163). Deep confidence, on this account, is conditional on the implementation of 'cures', 'cultures', or 'regimens' for the rectification of our corrupted minds.

The conditional status of deep confidence has implications for the conception of what the virtue of epistemic humility will require of an agent—for instance, does it require that we trust in 'grace' or in our 'natural light of reason' or rather undertake 'regimens of the mind'? I want to show, in the remainder of the paper, how nineteenth century philosophy offers us a diversity of rival metaphysical visions and their associated conceptions of epistemic humility.

## Doctrines of Deep Confidence: Comte, Peirce, and Spencer

The nineteenth century was philosophically heir to a rich heritage of developments – early modern disputes between 'rationalists' and 'empiricists', Absolute Idealism and its critics, and others. I will not attempt a survey of these developments and their relation to my theme of deep confidence. Instead, I offer three examples of nineteenth century philosophies that include doctrines of deep confidence.

The first is 'positivism', today especially associated with the writings of Auguste Comte. Its deep confidence was placed in 'the progressive growth of the human mind', guaranteed by the inevitable development of 'our principal conceptions' of knowledge through the famous 'Three Stages' – the theological, metaphysical, and scientific (or 'positive'). Such deep confidence runs through Comte's articulation of positivism, most obviously in the Law of Stage's assertion of 'general progress'. But it is there, too, in his emphases on our increasing proficiency in the 'vast intellectual operation' of organised science and the efficacy guaranteed by its 'unity of method'. Comte also, however, notes various conditional constraints on this confidence: it is 'rash to hope, even in the most distant future', that positive science could reduce the number of 'general laws ... to a single one' – a single theoretical description of the world – an epistemic ambition forever beyond our reach. Comte also warns that embrace of narrowly utilitarian conceptions of science would ensure we never surpass the 'present condition of mental development' (1988, 1, 2, 22, 30, 32, 41). In Comte's writings, then, is an account of the conditional status of our deep confidence, of a sort that can provide the basis for a positivist conception of epistemic humility.

A second nineteenth century development, also sympathetic to science, is pragmatism, especially as developed by Charles Sanders Peirce. It is 'certainly best' if our beliefs are suitably 'fixed', against the 'irritation of doubt', so that they can 'truly guide our actions' and so 'satisfy our desires'. Since our actions and beliefs are situated within a social world, Peirce characterises 'inquiry' in terms of social-epistemic practices for the 'fixation of belief', of which, in the modern world, science evidently claims 'the most wonderful triumphs in the way of settling opinion' (1992, 114, 120). Yet the main object of deep confidence is not science, but *logic*, defined as the 'art of reasoning', able to critically 'fix' beliefs. But such 'logicality' cannot be confined to solitary thinkers, lest criticism become 'endless', perpetuating doubt. Peirce therefore urges deep confidence in 'logicality', on condition that it 'must embrace the whole community', since public systems of critical discourse provide checks unavailable to isolated reasoning. Confidence in science is justified by virtue of its status, within modern society, as the best available means for realising the ideal of 'logicality' embedded in a 'whole community' (1992, 149). A further object of deep confidence is Peirce's vision of the 'conception of reality' that will emerge as 'the ultimate result of inquiry'. The 'social principle ... rooted intrinsically in logic' is therefore related to a broader epistemic ambition of the convergence of socialised, 'logical' inquiry (1998, 149).

The cases of positivism and pragmatism indicate forms of deep confidence in the social and epistemic fruits of science familiar to later audiences. But other nineteenth century philosophies have very different tastes – systematic and metaphysical in a way that later thinkers find exotic or grandiose. The 'synthetic' philosophy of Herbert Spencer, a giant of Victorian intellectual life, aspired to provide a 'statement of the ultimate principles discernible throughout all manifestations of the Absolute'. An evolutionary current flows through Spencer's vision of a metaphysical reality, manifesting itself at every level of reality – inorganic, organic, and 'super-organic', or sociopolitical – according to a law of 'evolution and dissolution'. Landscapes, creatures, societies, and systems of ideas gradually move from 'confused simplicity' to 'distinct complexity', as described by the various special sciences (1867, xi, 438). Biology and history, for instance, articulate particular manifestations of this cosmic law – the evolution of creatures from simple to complex forms, say, or of societies from the 'primitive' to the technological.

Spencer's deep confidence was in social and moral progress, guaranteed by our increasingly sophisticated understanding of the 'law of evolution and dissolution', through 'synthetic' philosophy. The culmination is modern liberal and technological cultures, like ours, that represent the most advanced realisation of the values of liberality and individualism. But Spencer places strict limits on the depth of our epistemic confidence: since the special sciences focus on specific principles and phenomena, the Absolute necessarily 'transcends' both our 'knowledge [and] conception'. Attempts to investigate it will find us 'continually thrown back', forcing a humbling, 'deepened conviction' that understanding the Absolute is an 'impossibility' – indeed, it is, therefore, also 'the Unknowable'. This rules out the possibility of deep confidence in the knowability of the nature of reality, something that Spencer is happy to accept. 'Our duty' is to conduct ourselves, 'with all humility', according to the 'established limits of our intelligence' – relative to the Unknowable – and failure to do so constitutes 'radical vice' (1867, ix, 545, 113, 108).

These nineteenth centuries philosophies went on to enjoy very different fortunes. Positivism and pragmatism resonated within a culture defined by the sciences and so fared well into the twentieth century. But Spencer's 'synthetic' philosophy, with its brooding talk of the 'Unknowable', sat badly within an increasingly epistemically confident culture. The result was the entrenchment, by at least the late 1860s, of a set of visions of our epistemic situation within the wider order of things that encouraged a deep epistemic confidence. Such confidence manifested itself in the ambitions that were increasingly embraced by the representatives of leading currents in philosophy and science – a newly ambitious spirit that was most vividly clear in the writings of a group now known as the scientific naturalists.

## Scientific Naturalism and 'Institutionalised Humility'

Conceptions of epistemic humility are defined relative to background metaphysical visions or worldviews. As objects of deep confidence change, so do prevailing ideas of what constitutes epistemically humble conduct. The striking epistemic confidence of nineteenth century European culture was a product of many factors, not least the rapid development of science. Indeed, by the 1870s, an increasingly robust confidence in science was being championed by a group of mainly British scientists, educators, and writers – advocates of a new picture of the world, to be provided by science and made the basis of significant educational and social reform (see Barton 1998).

Perhaps the most vociferous of the 'scientific naturalists' was T.H. Huxley – biologist and, later, 'Darwin's bulldog' – who declared that 'the order of nature is ascertainable by our faculties to an extent which is practically unlimited', specifically through scientific methods, hence the need for a 'complete and thorough scientific culture' (1872, 145, 62). A few decades earlier, such confidence could only have seemed exaggerated and hubristic, but a new confidence in scientific enquiry made such proclamations not only acceptable, but, increasingly, compelling. W.K. Clifford, best known today for an essay on the ethics of belief, similarly insisted that there is, 'to every reasonable question', an 'intelligible answer', which 'either we or posterity may know through the exercise of scientific thought'. Since intelligible, reasonable enquiry is confined to those trained in scientific 'frames of mind', all else is derided as 'superstition'. Such willingness to assign to science an exclusive responsibility for all 'reasonable' enquiry is a clear mark of Clifford's deep confidence in the picture of the world it has a privileged power to render 'intelligible' (1879, 155).

If critics worried about a lack of humility in these claims, then ones by other scientific naturalists seemed even more hubristic. In the provocative 'Belfast Address', John Tyndall called for all 'religious theories, schemes, and systems' aspiring to 'cosmogony' to 'submit to [science's] control'. To achieve, this, a 'purely scientific' education is needed, to propagate scientific habits of thought, albeit ones sensitive to the 'Power which gives fullness and tone' to our existence – a sort of naturalistic, 'higher pantheism' (1874, 61, 36, 65). Such Romantic talk of a 'Power' fades, however, from the later scientific naturalistic writings. Karl Pearson, writing at the very end of the century, confidently dismissed Kant's and Spencer's talk of unknowable things-in-themselves as 'futile', insisting that science is 'daily subduing' the remaining 'mysteries' of the world. But that is no loss, for what is 'essential to good citizenship' is not religious mystery, but only the ideal of the 'scientific man', 'striv[ing] for self-elimination', humble and unselfish (1892, 72, 112). By the end of the nineteenth century, what counts as epistemic humility was being redefined: where once it entailed strict limits on human epistemic powers, now it was judged to be compatible with grand ambitions to know the order of reality – a depth of epistemic ambition that Spencer, and others would have rebuked as 'dogmatic' vice.

I argued that an epistemically humble person regulates their conduct, including their assertions and ambitions. But those of the scientific naturalists were hugely confident, even hubristic – as when Tyndall declares the 'impregnable position' of science (1897, 520) – even as they asserted their *humility*. Indeed, their writings are suffused with sincere professions of acute epistemic humility. Despite his bullishness, Huxley lauds scientists for their willingness to 'sit down before the facts', forsaking pride to 'give up every preconceived notion', and so 'follow humbly', like 'a little child' (1913, 235). Tyndall described a 'condition of success' in science as the 'willingness to abandon all preconceived notions', a spirit of 'self-renunciation' that characterises an 'earnest prosecutor of science' who has attained 'humility' (1867, 72, 73). Of course, such professions were not confined to scientific naturalists—Peirce also takes humility as the virtue of scientists who must foreswear 'presumptive choice of hypotheses', 'prefer the truth to their own interest', and so on – 'a true elevation of the soul' (1974, 314).

How could the scientific naturalists profess humility while simultaneously embracing such deeply, even hubristically, confident ambitions? Answering this requires appeal to the call for conceptions of humility to be understood relative to a metaphysical vision. It became popular, in the nineteenth and twentieth centuries, to perceive science as (in my terminology) the best means for systematically fulfilling confidence conditions – the confidence of agents in their trained epistemic capacities, say, or confidence in collectively organised communities of enquiry. Starting at the agential level, a trainee scientist gains new skills, knowledge, and self-trust by undergoing specialist training. The modern infrastructure of scientific education and apprenticeship enable agents to build up their confidence in themselves as epistemic agents – that being the first 'level' of humility. Moving to the collective level, the scientific establishment is a powerful infrastructure apt for shared enquiry: a complex structure of disciplines, material cultures, investigative technologies – a realisation of Huxley's vision of a collective 'reservoir of ascertained truth' (1893, 254), vindicating Peirce's faith in 'logicality', expressed across a 'whole community' (1986, 284). An acute sense of epistemic humility is therefore justified by science's distinctive status as an exemplar of what the sociologist of science Robert Merton called 'institutional humility' (1957, 646).

This explanation of the tension between the confidence and humility of the scientific naturalists runs into an obvious problem. It applies only to agential and collective confidence and so presupposes that deep confidence in science and its governing picture of the world is already warranted. If that picture is hubristic, then so are the ambitions and attitudes rooted in it. So, did the scientific naturalists warrant their picture of the world?

## Humility, Hubris, and 'Pictures of the World'

The question of whether the scientific naturalists were humble or hubristic depends on their warrant for the picture of the world in terms of which they measured their confidence. I want to suggest that they not only did not, but could not, warrant that picture: therefore appraisals of their virtuous or vicious character are significantly complicated.

Despite their deep confidence in the naturalistic picture of the world, the naturalists failed to provide compelling arguments in support of its truth. Huxley vigorously attacked forms of 'speculative philosophy' for relying on empirically unprovable claims. But he then cheerfully concedes that the 'postulates' of physical science, such as invariant, universal laws of nature, are 'neither self-evident nor … demonstrable' (1894, 61). Similarly, Tyndall candidly admits that science ultimately derives its 'motive force' from an 'ultra-scientific source'. This 'source' is an 'intellectual vision' that has 'feeling, not knowledge, for its base', including a vivid sense of the 'mystery of man's relation to the universe' (1870, 65 and 'Explanatory Note'). Confidence in a naturalistic picture of the world ultimately requires a process of conversion, rather than the provision of evidence or reasons.

A critic might reply that the rationality of that picture of the world can at the least be secured by the critical rejection of its rivals. Even if naturalists cannot prove their picture to be a true one, they can at least remove any alternatives through reasoned critique. But their writings unfortunately show little evidence of such critical efforts. Although rival visions are consistently acknowledged, they are usually either derogated – as 'speculative', say – or else passed over. Tyndall surveys several rival visions of the world, mainly various brands of German Idealism, that would check his deep confidence in science. But his only response is a breezy admission that it is 'by no means easy to combat such notions' (1874, 57). Similarly, Huxley chides Kant for failing to grasp that 'the laboratory is the fore-court of the temple of philosophy'. But this ignores the transcendental idealist strictures on the epistemic status of sense-experience (1971, 166). Although Kant's epistemological doctrines should not be accepted dogmatically, it was as reasonable then as now to expect Huxley to have provided arguments against them, rather than opting to beg the question against them.

Such failures make the professed humility of the scientific naturalists suspect. A virtuously humble person takes care to regulate their confidence by modifying their claims in the light of just criticisms and rival claims. Such well-regulated confidence seems not to be in evidence among these nineteenth century scientific naturalists; indeed, underlying their writings is – as George Levine (2014, 79) says of the Belfast Address – a 'supreme confidence joined to a sense of righteous superiority'. The story of how the scientific naturalists came to successfully entrench their 'intellectual vision' was, as Matthew Stanley documents, due not to successful argumentation against rivals, but to their 'taking control of scientific education in Britain', 'naturalising theistic concepts', and serendipitous 'shifts in religious life'. Indeed, so successful were these strategies, that, by the early twentieth century, naturalism had come to seem 'obvious and inevitable', fuelling a potent 'confidence' that 'scientific exploration' could successfully challenge all 'current unknowns' (2014, 243, 248, 266). Can we therefore charge them with *epistemic hubris*, the vice of those whose radical confidence not only *is* not, but *can* not, be warranted?

I suggest that there is no easy answer to that question. A clue to the reason for this is Tyndall's remark that, ultimately, confidence in science is rooted in an 'intellectual vision' – a picture or worldview. Considered relative to that vision, claims about the profound epistemic power of science are not hubristic, for the only obstacles to a description of reality are ones of time, energy, and investment, of a sort that the emerging scientific culture will vouchsafe. If so, hubris takes different forms than it would for those, like Kant or Spencer, whose visions of the world incorporate ineliminable constraints on deep confidence in science (for instance, there can be no science of 'the Unknowable'). We can only discuss epistemic humility and hubris with careful reference to the background metaphysical visions against which the nature and limits of our epistemic capacities are to be measured. There is no neutral conception of our epistemic situation within the world to which to appeal to judge what the putatively objective limits of our epistemic powers are – so perhaps the best that we can do is to carefully describe the background visions against which a given community made its judgments.

It is clear that, by the late nineteenth century, a scientific picture of the world was rapidly being entrenched, altering the sorts of deep confidence judged to be plausible and compelling. But its entrenchment was not a universal feature of all European cultures: while that picture quickly took hold in energetically industrial countries, like Germany and Great Britain, the situation was different elsewhere. In France, for instance, a vigorous discourse affirming the 'bankruptcy of science' emerged during the 1870s, motivated by a diverse set of worries—'a romantic reaction against industrialism', 'defence of traditional educational values', and acute criticisms of science's failure to promote material wellbeing while risking the erosion of 'the ethical framework upon which European civilization rested' (MacLeod 1982, 2). While across the Channel, British scientific naturalists were rhapsodising about the ceaseless merits of science's intellectual and social vision, French academics were warning that if science had not yet 'plunged into total bankruptcy', then it had 'at least had its credit shaken by its considerable failures' (Paul 1968, 306). Charting the emergence and diffusion of the cultural and intellectual developments that affected deep confidence in science must encompass careful historical analysis, of the sort begun by Richard Olson in his 2008 book, *Science and Scientism in Nineteenth-century Europe*.

Subsequent cultural and intellectual history was, of course, to witness the generally successful entrenchment of deep confidence in science as a privileged, progressive source of social advance and descriptions of the ultimate nature of reality. It is no coincidence that, by the late nineteenth century, rival visions less sympathetic to deep confidence in science either died out – as British and German Idealism did – or were radically amended, as when the neo-Kantians abandoned the core doctrine of unknowable things-in-themselves. In the course of the twentieth century, critical reactions against science continued, motivated by various philosophical, political, and cultural concerns and developments – postpostivist philosophies of science, feminist and postcolonial critiques, contemporary resistances to sciences judged to conflict with religious values, among others. All of these testify to historical patterns or rhythms to changes in conceptions of our epistemic situation and of the wider nature of reality, that play out, among other things, in the definition of what counts as epistemically humble or hubristic conduct and enquiry.

Certainly the idea that conceptions of epistemic humility are indexed to metaphysical visions becomes a theme of much of *fin-de-siècle* philosophical culture. In the early decades of the twentieth century, many of the main philosophical movements, especially phenomenology, began to argue that our experience and enquiry ultimately presupposes conditions of possibility that cannot be justified – 'ways of being in the world', a 'life-world', and so on. Such claims incorporated latent doctrines of humility, for they impose strict constraints on the scope and strength of our epistemic capacities. But it is in the later writings of another philosopher, Ludwig Wittgenstein, that these thoughts find their fullest expression. Reflecting on the increasingly confident scientistic culture that was developing around the 1920s, Wittgenstein began to argue that science presupposes a *Weltbild*, an 'inherited background' that acted as the 'matter-of-course foundation' for our practices of 'enquiring and asserting', against which we 'distinguish true and false'. But a picture cannot, argued Wittgenstein, be proven or refuted, since any practices or criteria of proof and refutation would be drawn from the very picture whose legitimacy is being questioned (1967, §§94, 167; see Cooper 2017). Such remarks suggest that the most we can do is describe the naturalists' picture of the world and ask whether it is justified on its own terms, or confining criticism to other criteria – moral, say (see Kidd 2017). As Edward Craig puts it, a *Weltbild* – defined as a 'very general picture of the real' – can at best be 'articulated', by describing historically and philosophically their origins and development, but not definitively 'proven' (1987, 2).

It is not necessary to take a stand on these philosophical issues to appreciate that they reflect a new sort of epistemic humility. This new sort was rooted in anxieties about the possibility of justifying the very deep confidence in the scientific picture of the world that we have inherited, a set of anxieties that was the product of a particular and contingent cultural and intellectual history. In the course of that history, the concept of epistemic humility and its status as a virtue has changed, often quite dramatically, and continues to do so, so the historical story is on-going (and a very rich account of the history of epistemic humility is offered by Cooper 2002, chs. 2–6).

## Conclusions

This chapter aimed to show that at least some epistemic virtues and vices must be conceived with reference to changing background conceptions of the nature of reality, and sketched an account of a recent episode in the history of one such virtue: 'epistemic humility'. That history continued, of course, throughout the twentieth century, when debates about epistemic confidence in science continued unabated. Certain forms of constraint on that deep confidence were to fade away, even as others appeared to take their place. Few scientific realists today worry about postlapsarian limits to deep confidence in science, but instead reflect on the epistemic implications of its historical contingency (Kidd 2016b). Is talk of a comprehensive scientific description of the world impossibly hubristic, or an ambition compatible with a real sense of epistemic humility? If I am right, this is a debate about the conceptions of humility available in cultures where a naturalistic picture of the world prevails. But that is another episode in the history of the virtue of epistemic humility in science for another time.

**Acknowledgements** I am grateful to Chiara Ambrosio, David E. Cooper, colleagues at Durham, the editors, and the participants at the conference from which this volume emerged for helpful comments, encouragement, and discussion.

## References

Barton, Ruth. 1998. "Huxley, Lubbock, and half a dozen others": Professionals and gentlemen in the formation of the X Club, 1851–1864. *Isis* 89: 410–444.

Blumenberg, Hans. 1983. *The legitimacy of the modern age*. Cambridge, MA: MIT Press.

Clifford, William Kingdon. 1879. *Lectures and essays*, vol. 1, ed. Leslie Stephen and Frederick Pollard. London: Macmillan and Co.

Comte, Auguste. 1988. *Introduction to positive philosophy*, ed. and trans. Frederick Ferré. Indianapolis: Hackett.

Cooper, David E. 2002. *The measure of things: Humanism, humility, and mystery*. Oxford: Clarendon Press.

———. 2017. Superstition, science, and life. In *Wittgenstein and scientism*, ed. Jonathan Beale and Ian James Kidd, 28–38. New York: Routledge.

Corneanu, Sorana. 2011. *Regimens of the mind: Boyle, Locke, and the early modern cultura animi tradition*. Chicago: University of Chicago Press.

Craig, Edward. 1987. *The mind of God and the works of man*. Oxford: Clarendon Press.

Daston, Lorraine, and Peter Galison. 2007. *Objectivity*. New York: Zone Books.

Harrison, Peter. 2007. *The fall of man and the foundations of science*. Cambridge: Cambridge University Press.

Huxley, Thomas Henry. 1872. *Lay sermons, addresses, and reviews*. New York: D. Applegate and Company.

———. 1893. *Essays*. London: Macmillan.

———. 1894. *Methods and results: Essays*. London: Macmillan and Co..

———. 1913. *Life and letters of Thomas Henry Huxley*, vol. 1, ed. Leonard Huxley. London: Macmillan and Company.

———. 1971. *T.H. Huxley on education: A selection from his writings*, ed. Cyril Bibby. Cambridge: Cambridge University Press.

Kidd, Ian James. 2015. Educating for intellectual humility. In *Intellectual virtues and education: Essays in applied virtue epistemology*, ed. Jason Baehr, 54–70. London: Routledge.

———. 2016a. Intellectual humility, confidence, and argumentation. *Topoi* 35: 395–402.

———. 2016b. Inevitability, contingency, and epistemic humility. *Studies in History and Philosophy of Science* 55: 12–19.

———. 2017. Reawakening to wonder: Wittgenstein, Feyerabend, and scientism. In *Wittgenstein and scientism*, ed. Jonathan Beale and Ian James Kidd, 101–115. London: Routledge.

Kidd, M.S. *Deep conceptions of epistemic vice*. Unpublished paper.

Levine, George. 2014. Paradox: The art of scientific naturalism. In *Victorian scientific naturalism: Community, identity, continuity*, ed. Bernard Lightman and Gowan Dawson, 79–100. Chicago: University of Chicago Press.

Lloyd, Genevieve. 2000. Feminism in history of philosophy: Appropriating the past. In *The Cambridge companion to feminism in philosophy*, ed. Miranda Fricker and Jennifer Hornsby, 245–263. Cambridge: Cambridge University Press.

Macleod, Roy. 1982. The 'bankruptcy of science' debate: The creed of science and its critics, 1885–1900. *Science, Technology, and Human Values* 7 (4): 2–15.

Merton, Robert. 1957. Priorities in scientific discovery: A chapter in the sociology of science. *American Sociological Review* 22: 635–659.

Montaigne, Michel de. 1991. *Essays*, ed. and trans. M.A. Screech. London: Penguin.

Olson, Richard. 2008. *Science and scientism in nineteenth-century Europe*. Urbana: University of Illinois Press.
Pardue, Stephen T. 2013. *The mind of Christ: Humility and the intellect in early Christian theology*. London: Bloomsbury.
Paul, Harry W. 1968. The debate over the bankruptcy of science in 1895. *French Historical Studies* 5: 299–327.
Peirce, Charles Sanders. 1974. *Collected papers*, 2 vols., ed. Charles Hartshorne and Paul Weiss. Cambridge, MA: Belknap Press.
———. 1986. Writings of C.S. Peirce: A chronological edition, vol. 3. 1872–1878. Bloomington: Indiana University Press.
———. 1992. The essential Peirce: Selected philosophical writings: vol. 1, 1867–1893, e. Nathan Houser and Christian Kloesel. Bloomington: Indiana University Press.
———. 1998. *The essential Peirce: Selected philosophical writings*, vol. 2, 1893–1913, ed. Peirce Edition Project. Bloomington: Indiana University Press.
Roberts, Robert C., and W. Jay Wood. 2007. *Intellectual virtues: An essay in regulative epistemology*. Oxford: Clarendon Press.
Shapin, Steven. 2008. *The scientific life: A moral history of a late modern vocation*. Chicago: University of Chicago Press.
Spencer, Herbert. 1867. *First principles*. London: Williams and Norgate.
Stanley, Matthew. 2014. *Huxley's church and Maxwell's demon: From theistic science to naturalistic science*. Chicago: University of Chicago Press.
Tyndall, John. 1867. On the study of physics. In *The culture demanded by modern life: A series of addresses and arguments on the claims of scientific education*, ed. Anon. New York: D. Appleton and Company.
———. 1874. *Address delivered before the British Association Assembled at Belfast*. London: Green and Co..
———. 1897. *Fragments of science: A series of detached essays, addresses, and reviews*. Vol. 1. New York: D. Appleton and Company.
Wittgenstein, Ludwig. 1967. *On certainty*, ed. G.E.M. Anscombe and G.H. von Wright. Trans. Dennis Paul and G.E.M. Anscombe. Oxford: Basil Blackwell.
Zagzebski, Linda. 1996. *Virtues of the Mind: An Inquiry into the Nature of Knowledge and the Ethical Foundations of Knowledge*. Cambridge: Cambridge University Press.

# "Broken Symmetry": Physics, Aesthetics, and Moral Virtue in Nuclear Age America

**Jessica Wang**

**Abstract** During a century of high modernist confidence from roughly the last third of the nineteenth century onwards, scientists and other proponents of scientific investigation frequently claimed that scientific inquiry embodied both epistemic and moral virtues. This essay examines the lives and careers of two American physicists, Merle A. Tuve and Robert R. Wilson, in order to explore the history of the pure science ideal and the centrality of moral virtue to scientific identity in the mid-twentieth century. For Tuve and Wilson, belief in the humanistic qualities of science and the virtues—both epistemic and moral—of austerity and contemplative isolation, as well as an instinctive disdain for post-World War II Big Science despite the heavy involvement of both men in large-scale research projects, defined their faith in science as a morally virtuous activity. In a cold war era of military patronage, in which political and institutional trends seemed headed increasingly in the direction of instrumentalist expectation rather than transcendent humanism, Tuve's and Wilson's insistence on the moral virtues of scientific inquiry marked a larger struggle at work over the cultural meaning and status of physics, the nature and values of American culture more generally, and even over modernity itself.

**Keywords** Epistemic virtues • Moral virtues • Big science • Cold War science • Pure science • Nuclear physics • High energy physics • Particle physics • History of physics • Merle A. Tuve • Robert R. Wilson

In the fall of 1945, less than 3 months after the atomic bombings of Hiroshima and Nagasaki, J. Robert Oppenheimer delivered an address on the occasion of his departure from Los Alamos. He spoke before a gathering of some 500 members of the Association of Los Alamos Scientists, one of the local organizations of former Manhattan Project personnel engaged in an early postwar frenzy of political

J. Wang (✉)
University of British Columbia, Vancouver, BC, Canada
e-mail: jessica.wang@ubc.ca

 
J. van Dongen, H. Paul (eds.), *Epistemic Virtues in the Sciences and the Humanities*, Boston Studies in the Philosophy and History of Science 321,
https://doi.org/10.1007/978-3-319-48893-6_3

mobilization to promote nuclear arms control and head off what many scientists feared would be a devastating future of all-out nuclear destruction. In his speech, Oppenheimer famously described the inevitability of the wartime atomic project in terms of the inexorable nature of scientific discovery. "The reason that we did this job," he explained, "is because it was an organic necessity." Oppenheimer elaborated, "If you are a scientist you cannot stop such a thing. If you are a scientist you believe that it is good to find out how the world works; that it is good to find out what the realities are; that it is good to turn over to mankind at large the greatest possible power to control the world and to deal with it according to its lights and its values" (Oppenheimer 1945).[1]

Oppenheimer's words brought a familiar juxtaposition of the epistemic and moral virtue of scientific inquiry into an uneasy tension. Historically, epistemic and moral virtue went hand-in-hand in the modern world: scientific thought embodied epistemic virtue in its capacity to obtain knowledge about the natural world, and it connoted moral virtue because of the seemingly self-evident principle that, as Oppenheimer stated it, "it is good to find out." That latter assertion, however, rang increasingly hollow in a twentieth century era of total warfare, in which nuclear weapons now offered the potential to wipe out human achievement, if not all humanity itself. In the somber mood of the immediate postwar period, in which former Manhattan Project scientists were acutely conscious of how their unwavering wartime efforts had yielded human access to unprecedented powers of annihilation, more than a few members of the audience must have flinched inwardly and wondered whether it was necessarily all that good to find out and to know.

As the nuclear age dilemma outlined in Oppenheimer's farewell address indicates, no discussion of epistemic virtues can avoid for very long questions about moral virtue in science. The editors of this volume have already emphasized that specific virtues are not solely epistemic in nature, and a given character trait, such as honesty, humility, or objectivity, can possess epistemic qualities while also serving as a technology for the cultivation of individual scientific selfhood, or as a resource for manifesting community values and a common culture within the scientific enterprise (see the Introduction). Objectivity, for example, developed historically as both a means of obtaining reliable knowledge and a basis for creating a particular kind of self-disciplined, ethically virtuous personhood (Daston and Galison 2007: 39–41). In the process, objectivity also propelled science as part of a more widely dispersed cultural project of modernity, progress, and the advancement

[1] Oppenheimer's address deserves a more sustained analysis than is possible here, one that would explore fully the context behind the oft-quoted excerpt that I refer to above. His Los Alamos address highlighted the incompatibility between secrecy and science, to the point of declaring that information control "strikes at the very root of what science is" and threatened its very existence. Just weeks earlier, however, and to the chagrin of many Manhattan Project scientists, Oppenheimer had come out in favor of the May-Johnson bill, an early proposal that would have placed atomic energy in the United States under military control with strict secrecy requirements (Oppenheimer 1945; Smith 1965: 140–43). Oppenheimer's defense of science and the "organic necessity" of the bomb also did not absolve scientists of responsibility, although his conclusion refused to modify his affirmation of scientific knowledge as an unqualified good.

of human possibility (Herzig 2005; Shapin 2008). But such claims about the multifaceted virtues of science never went uncontested. The age-old idea of hubris underscored the price exacted for daring to know, while by 1945 a modern era of industry and warfare had already produced decades of commentary about the extent to which new scientific and technological capacities continually outpaced the development of human wisdom.

The nuclear age, however, drove a still more powerful wedge between the epistemic and moral value of scientific inquiry. That Oppenheimer in November 1945 felt compelled to defend science itself and insist that the quest for knowledge constituted an inherently moral activity suggests the extent to which the atomic bomb's wartime use challenged standard assumptions about the general beneficence and worthiness of knowledge acquisition as an end in and of itself. The guiding principle that "it is good to find out" no longer appeared axiomatic now that high science and technology had placed human existence in jeopardy. If the epistemic qualities of science no longer made self-evident the value of doing science, then what did?

This essay examines how American physicists Merle A. Tuve and Robert R. Wilson attempted to answer this question throughout their careers, particularly in the shadow of the nuclear age and in a postwar order that tied scientific research ever more tightly to the military-industrial complex. They responded with concerted efforts to articulate why science constituted a morally virtuous activity in an era when epistemic qualities alone could not justify the value of the scientific life. My title takes its inspiration from Robert R. Wilson's 1978 sculpture, "Broken Symmetry," which graces the Pine Street entrance of the Fermi National Accelerator Laboratory, located just outside Batavia, Illinois, with a series of three curved supports that meet to form an arch over the roadway (Figs. 1 and 2). When viewed directly from below or above, the work appears perfectly symmetrical, but all other angles reveal its fundamental asymmetry. "Broken Symmetry" conveys various references and meanings. At the most prosaic level, it calls attention to symmetry breaking and its centrality in elementary particle physics. In a slightly more figurative interpretation, Leon M. Lederman, Wilson's successor as director of Fermilab, once suggested that the sculpture referred to physicists' belief in "a hidden symmetry in what appears to be a very asymmetrical universe" (Lederman 1993: 26). For Wilson himself, the curved arches of the piece formed a visual reference to the flying buttresses that both ornament and support late medieval cathedrals such as the Beauvais Cathedral, which Wilson visited in 1965. In subsequent years, Wilson cited the Beauvais Cathedral as an inspiration for the design of the Central Laboratory at Fermilab, and he expressed the aesthetic hope that particle accelerators, like Gothic cathedrals, could "express the aspirations and spirituality of their age" (Wilson 1987; Hoddeson et al. 2008: 66). In political terms, the sculpture represents a call for peace and the beating of swords into plowshares, since it was constructed out of deck plating from a retired aircraft carrier, which also provided radiation shielding within Fermilab's research facilities (*FermiNews* 1978). The artwork offers a cultural commentary and perspective on scientific reasoning as well:

**Fig. 1** "Broken Symmetry," side view (Photograph by Reidar Hahn, 1998, Courtesy of Fermi National Accelerator Laboratory)

**Fig. 2** "Broken Symmetry," view from below (Photograph by Fred Ulrich, 1998, Courtesy of Fermi National Accelerator Laboratory)

"The piece symbolizes the mission of Fermilab and, indeed, of the scientific process itself; to look at problems from all sides, and to discover truth by so doing" (*Proton News* n.d.).

Taken in combination, these different aspects of "Broken Symmetry" also speak to Wilson's desire to combine aesthetic values and ethical commitments with the political and institutional circumstances of big science in cold war America. More broadly, this effort on Wilson's part exemplified the faith of many elite physicists in the post-World War II era in an ideal of pure science that emphasized the moral and aesthetic dimensions of science, held that scientific inquiry was a humanistic enterprise that built character and expressed high cultural values, and viewed basic research as a worthy endeavor devoted to the pursuit of knowledge for its own sake. History of science, as part of its early disciplinary identity, also perpetuated an account of science as the acme of human achievement and the foremost measure of civilizational progress (Herzig 2005; Shapin 2008; Dennis 1997). In recent years, most scholars in the history of science and science studies, for understandable reasons, have preferred to dismiss this ideal of pure science as rhetorical window dressing that masked darker material and political realities. A generation of scholarship has associated cold war scientific research with the priorities and agendas of the national security state, and a large body of work has exposed the different ways in which all manner of research projects in the natural and social sciences that purported to be value-free, pure science all too frequently reflected cold war patrons' specific needs and objectives.[2] Yet it is still worth contemplating the history of the pure science ideal, because we will not understand much about scientists' self-conception or the cultural status of science in mid-twentieth century America without taking the notion seriously.

Wilson's meditation on perspective, along with this volume's overall emphasis on the multifaceted nature of virtues, invites us to remain sensitive to multiplicity and contradiction in our analyses of virtue epistemology and the cultural rationales for doing science. My own interest here lies in trying to understand and make meaning of humanistic and ethical commitments in mid-twentieth century physics while also recognizing the extent to which the cold war political economy of science and the intricate ties between physics and the postwar defense establishment belied the very notion of pure science. If we "look at problems from all sides," we might be able to see both the aspiration to virtuous creativity and the immersion of American physics in military problems and the exigencies of global ideological struggle as part of the same structure and enterprise, rather than simply dismissing the former as rhetorical and conceptual fluff designed to obscure underlying realities. The central issue here is not the normative question of whether or not physics is, or can be, a humanistic activity that embodies moral virtues, but a set of historical questions: Why, in the middle decades of the twentieth century, did elite physicists talk about their work as a humanistic endeavor and a search for knowledge for its own sake? Why was it plausible, and even necessary, to think about science in this way? And what does this history have to tell us about our own time period and the limits of an

[2] As an entry point into this scholarship, see Forman 1989; Leslie 1993; Simpson 1999.

instrumentalism that has arguably come to define the value of science in twenty-first century America?

Tuve's and Wilson's lives provide ideal opportunities for addressing these questions, because both physicists, although roughly an academic generation apart in time, strived in similar ways to sustain a humanistic rationale for their discipline in the face of the jarringly destructive mélange of physics and state power from World War II onwards.[3] Both shared a belief in science as a humanistic activity, a faith in the virtues of austerity and contemplative isolation, and a disdain for an era of Big Science and team research despite their own participation in such large-scale scientific and technological ventures. Both also sought, and at some deeply personal level needed, a social rationale for science beyond the relevance to weapons systems and other requirements of the cold war national security state that had helped make epistemic virtue an increasingly unsatisfying basis for the moral virtue of science. Physics now needed to look beyond the presumed value of knowledge-making for its moral foundations.

For Tuve, a fundamental tension defined his scientific life throughout the postwar period. On the one hand, he constantly articulated and defended an ideal of pure science that celebrated asceticism, austerity, and the virtues of scientific research as an isolated and contemplative search for truth and beauty. On the other hand, he remained embedded within both the national security state and big science, two institutions that he simultaneously decried yet, however reluctantly, embraced. This tension was not simply a case of ideology masking reality, but, for Tuve, a profound struggle related to the cultural purpose and legitimacy of both science and scientist.

Tuve launched his career in the 1920s and 1930s as an American-born, American-trained physicist from a midwest, Protestant background. He first studied physics at the University of Minnesota before earning his doctorate at Johns Hopkins University, for research under the direction of Gregory Breit at the Department of Terrestrial Magnetism (DTM) of the Carnegie Institution of Washington (CIW), where Tuve worked on using radio waves to study the ionosphere. He subsequently joined the DTM's staff and shifted his research interests toward nuclear physics. In collaboration with his DTM colleagues, and with the aid of the CIW's state-of-the-art Van de Graaf generator, Tuve conducted path-breaking experiments on proton-proton interactions in the 1930s. Then, like virtually all American physicists, he turned to military projects during World War II. Tuve spent the war at Johns Hopkins University's Applied Physics Laboratory, where he directed the project that took the proximity fuze from development to production. Afterwards, he returned to the Carnegie Institution of Washington to take the reins as director of the Department of Terrestrial Magnetism.

The war and its aftermath left Tuve suspended between conflicting desires and responsibilities: a desire to retreat to a world of science insulated from public affairs, an attraction toward scientific problems that did not allow such retreat, a set of ambitions that also militated against an isolated existence in the laboratory, and a

[3] A more detailed and extensive account of my work on Tuve appears in Wang 2012.

sense of obligation that, particularly after the start of the Korean war, required engagement with the corridors of power. In response to this complex set of contradictory commitments, throughout the postwar years, Tuve attempted to construct, articulate, and disseminate a particular set of notions about scientific identity, centered on the ideal of pure science and tied to a morally virtuous scientific persona. In part, that response consisted of public statements about the moral, aesthetic, and humanistic virtues that he identified with scientific inquiry and the persona of the scientist. For example, in March 1947 at the Westinghouse Sixth Annual Science Talent Search, he told the audience that the significance of science lay in its humanistic purpose "to orient man in his relationship to the world in which he finds himself, including his relationship to fellow men, to [the] past, to [the] future and to the infinite and awe inspiring mysteries by which he finds himself surrounded." Such a worldview imbued the scientist with moral qualities: "Humility[,] sincerity, honesty and dedication to something greater than himself—these are virtues of the scientific life." The "great structures of human knowledge and understanding" also carried "aesthetic values" that were "as vivid and as moving as any in the realms of art and music." Science situated human experience within a larger universe, shaped the moral being of the scientist, and reached to the depths of the soul with its power to imbue feeling. No mere exercise in reason and application, science served the higher goal of cultivating the human spirit (Tuve 1947a).

In addition to such encomiums to science—and also public statements that openly condemned cold war funding trends that he thought damaging to true scholarship—Tuve tried to enact these ideals as a lived practice. He saw the CIW as ideally situated to pursue small-scale, cutting-edge problems that required a level of creativity largely absent in the brave new postwar world of fancy equipment, cumbersome research teams, and seemingly endless supplies of cash that defined big science. As part of a long tradition that identified the scientific life with asceticism, Tuve tied the austerity of the DTM's postwar research environment not just to the moral virtue of self-denial and the creativity that he thought came from small-scale research, but also to the freedom of inquiry possessed at an independent research foundation with its own endowment. Such an institution created the possibility of virtuous isolation in support of scholarly contemplation, a particular necessity in an unstable world of growing international strife. Tuve's conscious postwar retreat from nuclear physics, once a "philosophical field" and now "a welter of power politics," directly reflected that effort to find a principled refuge in solitude (Tuve 1947b). In addition, Tuve tried to simplify his institutional commitments in order to stay out of the evolving cold war political economy of science. In the years prior to the Korean war, he turned down a significant number of prominent advisory positions within the defense establishment, and he sometimes did so in harsh, even overwrought, terms. As Tuve told the Research and Development Board in the spring of 1950: "the proposal [to join the Board] does not arouse in me any spark of challenge or spontaneous response, but only looms more and more as a burden and a painful task.... I cannot spend a third or a half of my time on work for the RDB without wrecking most of what is valuable in my life....The demands of research are not easy to meet at my age and with the interruptions we have all had, but I am sure the goal is worth while. Taking the RDB assignment now would place it permanently out of my reach" (Tuve 1950). The language mattered—his lack of any "spark" of interest, his sense of the RDB post as a "burden" and "painful," his belief that acceptance would require "wrecking" the aspects of his life that he treasured the most—such emotional terms underscored a deep-seated and intensely personal effort to stay out of the cold war partnership between science and the military.

At the same time, Tuve's postwar career did not offer a simple alternative outside the cold war political economy of science, and even with his powerful qualms, he could not (or did not) extricate himself entirely from science-military relations. His early postwar research program in geophysics required frequent collaboration with the Navy, as a supplier of TNT and explosives experts to set off detonations that would generate waves yielding information about the structure and contours of the earth's crust. After the start of the Korean war (or, rather, the escalation of Korea's ongoing civil war in June 1950 that Americans use to date the onset of the conflict), Tuve also consulted frequently for the military, despite his antipathy towards such work. Drawn by a sense of civic obligation, he plunged back into a world of science-military relations that he had rejected just a few months earlier, and he never again contemplated the kind of self-imposed separation from the national security state that he had attempted in the early post-World War II years (Wang 2012: 353–57).

In part, Tuve's struggle reflected difficult choices for a generation of scientists that also believed in civic duty and public service, which did not make opting out for the sake of moral purity a simple matter. More germane to this volume's concerns with the relationship between the virtues and scientific selfhood, Tuve's sense of conflict matters as a way to understand scientific identity and its larger implications. Most accounts of scientific identity, such as Lorraine Daston and Peter Galison's *Objectivity*, tie scientific selfhood to questions about objectivity and its epistemic virtues. But there is more to scientific identity than the self-abnegating, self-denying effort to create "knowledge that bears no trace of the knower" (Daston and Galison 2007: 17). Tuve's emphasis on emotional and spiritual engagement prioritized the centrality of the affective self to the moral character of the scientific enterprise, and this insistence on the importance of moral virtue resonated with broader discourses about scientific identity in the United States during the period from the 1870s to the 1960s. Tuve represented just one example of a mid-century preoccupation, one that grew in prominence during the late 1930s and accelerated in the aftermath of Hiroshima, with the moral and ethical status of science and scientists. In academic settings, journals of opinion, committees of learned societies, sociological analyses, religious circles, and a wide variety of public forums, a broad range of figures—American scientists, philosophers, religious leaders, and public intellectuals—struggled to articulate the nature of both science as an activity and the people engaged in it (Wang 2012: 363–64).

In this context, Tuve offered the construct of an idealized scientific persona centered on a sense of imagination, the power of research to nurture and enrich the human soul, a scientific attitude defined by a "spirit of wonder," an acutely personal and emotional attachment to science, and dedication to the humanistic dimensions of science which saw science as "part of the liberalizing humanities" (Tuve 1958, for the quotations; Wang 2012: 365–70). This humanistic outlook also viewed dedication to science as an aesthetic pursuit, one that instilled devotion to a way of life that sought and idealized isolation in the search for knowledge. For Tuve, the "quiet research men" of science were the real movers and shakers who advanced knowledge and values, by pursuing private contemplation and eschewing the wheeling and dealing of policy entrepreneurship. The latter belonged instead to the lowly world of organization men who let personal ambition override concrete scientific achievement (Tuve n.d. [1959?]; and Tuve n.d. [1960?]).

The fabricated and fictive nature of Tuve's idealized scientific persona should not lead one to ignore the cultural work it attempted to perform. As Lorraine Daston and Otto Sibum have suggested, personae "are nonetheless real, in that they cause things to happen in the world," and their fictive qualities "may be mythologies not in the historian's sense of being false to facts, but in the anthropologist's sense of being true to essences" (Daston and Sibum 2003: 7–8). From the late nineteenth century to the 1960s, the image of pure science provided scientific practitioners with a built-in response to a number of disquieting trends: the challenge of industrial capitalism and a new culture of acquisitiveness to Protestant traditions that identified self-denial with moral uplift; the rise of engineering, and then industry-based science, and an expansion of employment opportunities that threatened to transform science from a calling into a job already in the early decades of the twentieth century; and the mid-twentieth century anxiety over the nuclear age, with its mixed messages of the inhuman implications of scientific discovery and the powerful possibilities of science as instrumentality (Herzig 2005: chapter 3; Kline 1995; Gilbert 1997: chapters 1, 3–4, and 12; Shapin 2008). To that extent, the assertion of science's distinctiveness, moral purpose, and inculcation of virtue in the scientific practitioner functioned as both a system of belief for the construction of identity as well as an anxious response to all of the trends and circumstances that suggested that science might, in reality, be none of these things.

Indeed, Tuve's own efforts to assert the value of scientific selfhood ultimately could not hold society at bay even within the contours of his own, private perception of being. The historical literature has documented clearly the frequent contradiction between claims to pure science and the material realities and practices of cold war science. For Tuve, the scientific life in the nuclear age undermined his very sense of self, particularly as he moved from an early postwar anticipation of a return to the "fun" of research in peacetime and the pursuit of scientific knowledge in its most humanistic and spiritually uplifting dimensions to an abject despair about the nuclear age and its dangers, in which grim international realities made it difficult to maintain faith in the scientific spirit (Wang 2012: 376–78). That despair, on at least one occasion, led Tuve to question his very sense of personhood. In August 1959, Bishop C. Gresham Marmion, chairman of the Episcopal Church's Joint Commission on "The Church in Human Affairs," asked Tuve to write a lay sermon offering a Christian perspective on the uses of nuclear energy. Tuve, a minister's son, agreed to the task but found himself paralyzed with doubts as he attempted repeated drafts without satisfaction. A complex tangle of feelings about physicists' hubris, the destructive legacy of nuclear physics, and his own sense of himself stood in the way.

He worried at length about the seeming arrogance of physicists claiming to offer wisdom and counsel when they themselves had created the technological means for humankind's annihilation. Above that, he felt keenly his own sense of personal inadequacy: "No man of my close acquaintance is this sure of himself or of his beliefs and his spiritual health. We all have shortcomings and lapses into helplessness, apathy and lethargy. I cannot picture that I am a truly whole person, not an especially pious one, and certainly must refuse to be singled out and put on a pedestal as a spiritual counselor" (Tuve 1959a). Not only was the physicist in the abstract unworthy of the pulpit, but Tuve saw himself as someone whose spiritual state, including a lack of full personhood, disqualified him from preaching to others.

The emotional nature of Tuve's response to the cold war order resonated strongly with the times. Within scholarly studies of emotion and the so-called "affective turn," one school of thought focuses on understanding emotional experience in social terms, as part of the sensibility of an era.[4] One perhaps cannot derive too much from a single person's emotional experience, but it is tempting to see in Tuve's despair a broader sensibility or structure of feeling that defined the cold war itself, in which a bleak emotional outlook matched the cold geopolitical conditions of the U.S.-Soviet conflict (Wang 2012: 384–86). Tuve was far from alone in experiencing the nuclear fear that shadowed everyday life in America in the postwar decades. In that context, Tuve's visceral experiences serve, at the very least, as another reminder of the emotional texture of the cold war era, particularly the pressures that shaped scientific identity and destabilized the assumption that "it is good to find out."

Robert R. Wilson's career offers an opportunity to place discussions of science and moral virtue in a different cold war context: that of the late 1960s and 1970s, a new era of budgetary constraint and decreased political support for American physics, as well as a time in which the pure science ideal as a form of identity was arguably itself in decline. David Kaiser has aptly termed this latter process "the suburbanization of American physics" to denote how physics increasingly went from being a calling to a job starting in the 1950s (Kaiser 2004). Meanwhile, by the early 1960s, the rapid increase in numbers of physicists combined with unsustainable growth rates in federal funding began to place significant constraints on basic research. As the National Research Council's Physics Survey Committee noted in 1966, the budgetary increases of 6–7% per year during the period from 1963 to 1965 (unimaginable largesse in American science today) could not match past annual growth rates of 20–29%, and basic physics required funding increases of 21% per year for the next four years to catch up to expectations (National Research Council Physics Survey Committee 1966: 88, 96). In this changed fiscal environment, physicists struggled to articulate a public rationale for the massive and expensive research projects that defined postwar big science, particularly in high energy physics. Three lines of reasoning appeared over and over again in the 1960s and early 1970s: first, a recapitulation of Vannevar Bush's classic argument from his

[4] For examples of analytical approaches that emphasize the social, collective dimensions of emotional experience, see Williams 1977; Stearns and Stearns 1985; Bourke 2003; Costigliola 2004; and Wickberg 2007.

1945 manifesto, *Science the Endless Frontier*, that basic research without immediate or obvious applications provided the foundation for future technological advances; second, a contention that basic science served valuable social and political goals, such as the advancement of national prestige or the promotion of international cooperation; and third, a plea for recognition of the cultural and humanistic value of particle physics.[5] Perhaps no other physicist expressed that third rationale more clearly and consistently than Robert R. Wilson.

Wilson earned his undergraduate and doctoral degrees at Berkeley in the 1930s, where he reveled in life with the cyclotron at Ernest O. Lawrence's Radiation Laboratory. He then took a position at Princeton, but World War II soon intervened, and he ended up at Los Alamos as a group leader on the Manhattan Project. Wilson found himself profoundly shaken by the destruction of Hiroshima and Nagasaki, and in the postwar years, he deliberately avoided any research that required a security clearance. He also became politically active as part of the atomic scientists' movement and joined the early post-World War II fight for civilian and international control of atomic energy (Wilson 1970a; Wilson 1970b; Hilts 1984: chapters 8–10).

In 1947, after a brief stint at Harvard, Wilson moved to Cornell University, which boasted the recently established Laboratory for Nuclear Studies under the leadership of Robert F. Bacher. Just a year later, Wilson became the laboratory's director, after Bacher left to become a commissioner on the Atomic Energy Commission, and Wilson took charge of the development of Cornell's synchrotron. During the late 1940s and the 1950s, Cornell offered somewhat of a refuge from cold war science. Although accelerator physics could not escape patronage from the Atomic Energy Commission, Wilson, as well as Philip Morrison, at least managed to avoid secret research and the most obviously militarized areas of physics (Schweber 1992; Schrecker 1986: 150–60). Wilson headed the laboratory for nearly twenty years, until he became Fermilab's first director in 1967. There he oversaw the building of the Main Ring accelerator during a difficult period of retrenchment in federal funding that marked an end to the heyday of American science and the era of unquestioned support for basic research.

Although Wilson was thirteen years younger than Merle Tuve, his career mirrored Tuve's in important respects. Both physicists shared a similar background as American-trained experimentalists of westerly and Protestant origins, whose introduction to accelerator physics came during the interwar years, the period which physicists later looked upon nostalgically as an age of "string and sealing wax." For Wilson, the joys of scrounging for equipment, tinkering in the laboratory, and risking the danger of burns, electrocution, fire, and other hazards all added to the delight and adventure of youthful scientific endeavors (Hilts 1984: 44–51). Early memories of working independently while being part of a small community absorbed by hands-on fiddling with a large, complex apparatus also contributed to a romanticized image of devotion to a life of science and a state of idealized creative isolation. Wilson later recalled the heady feeling of his undergraduate days at the Rad Lab: "I

---

[5] See, for example, Weinberg 1963; Yuan 1965; National Research Council Physics Survey Committee 1966: 88–89; Weisskopf 1970; and Rosen 1971.

was completely independent, and felt I was running my own laboratory, and my own research" (Wilson 1977: 164). He also venerated the protagonist of Sinclair Lewis' 1925 novel *Arrowsmith* and thought of the eponymous hero in terms that echoed Tuve's vision of creative isolation. Wilson later told journalist Philip J. Hilts, "That romantic idealization of a man dedicated to research made a deep impression upon me . . . a young and devoted medical scientist who, working long hours in solitude, finally experiences the ultimate exaltation that comes to any creative person. I was a lonely boy in Wyoming, and for some reason it was natural for me to relate to Lewis' hero … [I] could empathize with Arrowsmith on his lonely research frontier" (Hilts 1984: 43).[6]

Wilson spoke much the same language as Tuve when it came to the virtues of scientific solitude, the elevation of austerity to a moral value, the aesthetic and spiritual qualities of physics, and the tensions between hoped-for ideals and more sobering realities. In his 1968 Richtmyer lecture, for example, he lauded the "hope for isolation" and the moments "spent in lonely, almost monastic, meditation," even as he acknowledged as an illusion the idea that physics could separate itself from the political necessities of the nuclear age (Wilson 1968: 490). The virtues of austerity also marked Wilson's career at important junctures. As a young physicist during the Great Depression, at least as he recalled his thought process in retrospect, he linked the era's economic deprivations to the valor of physics as a career. As Wilson recalled to Spencer R. Weart in a 1977 interview, "My thinking was . . . that I wasn't going to get a job, no matter what I did. And if I was going to be unemployed, rather than being an unemployed engineer, an unemployed whatever, I was going to be an unemployed physicist. That was better" (Wilson 1977: 168). That willingness to embrace austerity served Wilson well decades later at Fermilab, which eschewed expensive, failsafe engineering in favor of building on the cheap, an approach that complemented the tight fiscal environment of the Vietnam era. Wilson justified the often primitive working conditions at Fermilab—poor climate control and water leakage in areas holding instruments, outdoor work spaces, a half-baked effort to use a trained ferret to clean out the accelerator's vacuum tubes—as part of an independent, self-reliant, pioneering scientific life, in which researchers roughed it on the frontier of physics (Hoddeson et al. 2008: chapters 6 and 7).[7] Wilson also shared Tuve's resistance to large team research, although as director of Fermilab he ultimately reconciled himself to the managerial role of accelerator builder, much as Tuve acceded to the age of big science when he pursued radio astronomy in the late 1950s (Wilson 1970a; Needell 1987).

---

[6] Similarly, when Spencer R. Weart once asked Wilson about his perception of Ernest O. Lawrence and whether he saw his mentor in the mode of Arrowsmith, Wilson immediately responded, "I didn't think of him as Arrowsmith, I thought of myself as Arrowsmith." Wilson 1977: 175.

[7] One should not idolize science on the cheap or underestimate its risks of failure. Fermilab's "magnet crisis" in the spring and summer of 1971, in which disastrous problems with the Main Ring's magnets led costs and delays to mount, suggested to Wilson's critics the tenuousness of the effort. Fortunately for Wilson, the Main Ring still managed to deliver its targeted energy level of 200 BeV in March 1972, within his promised timeline and budget (Hoddeson et al. 2008: 148–54; Hilts 1984: 94–98).

In addition, Wilson continued to invoke the old language of science as an aesthetic and spiritual quest, even as a younger generation perhaps saw science in different, more instrumentalist and careerist, terms. Frequent statements about the humanistic value of pure scientific research appeared in his public lectures, Fermilab's annual reports, and even in Wilson's Congressional testimony. As Wilson observed in his Richtmyer lecture, scientists needed to do more to convince the public and "persuade the humanists that there is real humanistic content and appeal" in physics and its exploration of nature (Wilson 1968: 493). Another typical formulation, in Fermilab's 1968 annual report, declared that "Pure science, the search for understanding, is as important for its effect on the minds of men as it is for its eventual contributions to his standard of living. Man's effort to achieve a better comprehension of the world in which he lives will continue to have a profound effect not only on his philosophy, not only his well-being, but also on his whole social organization" (Hoddeson et al. 2008: 113). In 1972, when questioned at a Congressional hearing about the expected practical applications of research at Fermilab, Wilson placed "cultural value" first and foremost and put science alongside "the other values of a great Nation which has a literature, has theatre, has poets and painters and sculptors, composers and musicians; those cultural things that make life richer" (Wilson 1972). He also told Philip J. Hilts, in language that Merle Tuve would have easily recognized, "Our work here [at Fermilab] is primarily spiritual.... We are concerned with the ultimate nature of matter" (Hilts 1984: 18).

Wilson's most distinctive contribution to this discourse about the ideal of pure science came in the form of his own work as an artist who specialized in sculpture and metalwork, and in his effort to create an aesthetic environment at Fermilab even amid the raw and frequently chaotic conditions there. By the mid-1960s, artistic commitments increasingly occupied Wilson's thoughts, and he was pondering whether to leave science altogether to devote himself full-time to sculpture (Hilts 1984: 83). He took drawing classes in Paris in the mid-1960s, and from French cathedrals he developed a desire, which he expressed frequently in subsequent years, to make accelerators into the cathedrals of the late twentieth century. As he noted in 1968, "cathedrals were technological constructions of great beauty and significance that expressed the aspirations and spirituality of their age," and Wilson wanted accelerators to serve that same cultural role in his own historical moment (Wilson 1968: 491–92, for the quotation; Hoddeson et al. 2008: 65–66; Jungk 1969: 216) (Fig. 3).

As part of that vision, Wilson incorporated aesthetics into the design of the entire Fermilab facility. "My fantasy of a utopian laboratory," Wilson later recalled, "clearly required a setting of environmental beauty, of architectural grandeur, of cultural splendor." He was determined to achieve his artistic goals despite Fermilab's parsimonious budget (Wilson 1987). As Lillian Hoddeson, Adrienne W. Kolb, and Catherine Westfall have noted, prairie restoration, in a reflection of both late 1960s environmentalism and Wilson's frequently invoked metaphor of the frontier, formed an important component of his "Renaissance naturalism." Wilson also hired an artist, Angela Gonzales, to serve as Fermilab's "aesthetic watchdog" and oversee site design, as well as artwork in the laboratory's publications and publicity materials

**Fig. 3** Reflection Pond, Hyperbolic Obelisk (Acqua Alle Funi) Sculpture and Wilson Hall (Photograph by Reidar Hahn, 2006, Courtesy of Fermi National Accelerator Laboratory)

(Hoddeson et al. 2008: 96, 121–22). Wilson himself took a personal interest in negotiating and judging architectural designs for Fermilab's main Central Laboratory, and to him, the resulting building (now known as Wilson Hall) shared a satisfying relationship and sense of perspective with the surrounding landscape reminiscent of the Beauvais Cathedral (Wilson 1987). Several of Wilson's major sculptures, including the striking hyperbolic obelisk "Acqua alle Funi," the "Mobius Strip" outside Ramsey Hall, and "Tractricious" (made in part from scrap tubing from the Tevatron magnets), along with "Broken Symmetry," adorn the grounds of Fermilab.

The question remains: what meaning, if any, does this discourse about the moral and aesthetic virtues of science add to the history of twentieth century science, particularly a political history of American science that has exposed the embeddedness of research in cold war defense needs, and what does it add to the history of epistemic virtues? In answer to the first part of the question, Tuve's and Wilson's experiences provide a reminder that, contrary to analyses that have emphasized scientists' cold war-era "false consciousness" about the purity of their research, physicists frequently retained an all too real sense of how the relationship between science and the state was reorienting their discipline towards instrumental and destructive ends.[8] Wilson himself, for example, recognized the limits of physicists' ability to shape the larger course of the nuclear age, as well as their place within the cold war order. His own postwar renunciation of secret research, he acknowledged in 1970, constituted

[8] Paul Forman famously held physicists to account for their "false consciousness" in Forman 1987: 228.

an imperfect response that was "good for my conscience," but that also "immediately reduced my effectiveness to do something about nuclear energy." Wilson explained, "My expertise soon became outdated: I had to watch my more conservative friends, usually working from within the government, give the kind of advice and exert the kind of political pressure that is based upon understanding."[9] After the American entry into the Korean War, Tuve followed the alternative strategy of returning to the corridors of power even as he continued to regret what he viewed as the cold war's corrosive distortion of scientific values. During the postwar decades, although physicists contributed to a few political victories, such as the establishment of civilian control of atomic energy, the founding of the National Science Foundation, and the ban on above-ground nuclear testing, Wilson viewed their overall results on arms control as a discouraging record of defeat (Wilson 1970b: 33). Moreover, Wilson concluded bitterly, physicists also completely overestimated their political influence. Although Manhattan Project scientists had hoped to determine the uses of the new and terrifying technology they had brought into being, ultimately "we were just like the slaves building the pyramids" (Hilts 1984: 79; see also Wilson 1970b: 33).

Tuve's and Wilson's struggle to find an appropriate balance between the competing demands of moral virtue and political responsibility also underscored the problematic nature of virtue in science, as the gap between epistemic virtue and its traditional moral counterpart widened during the post-World War II period. As Wilson's trenchant comments about the limits of physicists' political power indicated, physicists' version of science as a source of moral virtue was never able to match rival claims of nationalism, security, and national identity as public goods and virtues according to the cold war state's rendition of "it is good to find out." Partly, the problem was economic in nature: apart from the instrumentalism of the national security state and its demands for new technologies, nationalism provided the only other basis for supporting forms of physics so expensive and so immune to profit that only nation-states, or conglomerates of nation-states, could undertake the work. In that context, during the mid-twentieth century, the social or collective counterpart to the virtuous individual physicist became the nation-state's mission to demonstrate national prowess through monumental projects and the cultural diplomacy that they performed. In the United States, in addition to providing the basis of a militarized national security, massive technological projects—dams, high-energy particle accelerators, manned space programs—functioned in the twentieth century as demonstrations of what it meant to be a modern and powerful nation-state.[10] The

[9] Consider, for example, the ascendance of scientists such as Edward Teller and his allies, all staunch cold war hawks, within the Atomic Energy Commission in the 1950s. Indeed, President Dwight D. Eisenhower established the President's Science Advisory Committee (PSAC) in part to contain the power of the conservative scientists within the nuclear establishment. McGrath 2002: chapters 5–6.

[10] On dams as symbols of modernity and development, see Hughes 1989: chapter 8; and Ekbladh 2010. On space programs, the cold war, and the international politics of technological display, see McDougall 1985; and Smith 1983.

modern role of science and technology as means of reinforcing the collective virtue of "nation-ness" and national belonging, however unsavory its many offshoots, formed an alternative structure of values that advocates of the pure science ideal never fully reckoned with. Nor are such ideological functions of high modernist science and technology restricted to a cold war past. It is no accident, for example, that the post-Mao Chinese government, which seems to hew to a curiously old-fashioned version of modernity, has prioritized dam-building (especially the infamous Three Gorges Dam), a space program, and playing host to the Olympics, among other symbols of national prowess drawn from a mid-twentieth century repertoire. Not surprisingly, since 1989, China has also built several major particle accelerators, and particle physicists are looking to the PRC as the most likely state willing to build the next generation of supercolliders to supersede CERN's Large Hadron Collider (Overbye 2006; Gibney 2014).[11]

Meanwhile, in the United States, the cultural conditions that supported the pure science ideal slowly eroded. Over the longer term of the twentieth century, Wilson's effort to make Fermilab an aesthetic expression of a particular kind of scientific life came just as pure science itself lost its cultural cachet and authority. The pure science ideal originated from two distinctive historical contexts: a late-nineteenth century Protestantism that idealized self-discipline and self-sacrifice as the basis of a meaningful and moral existence (Herzig 2005); and a bourgeois German milieu, also from the late-nineteenth century, that valued a particular kind of cultured, educated, politically aware person engaged in a mission of self-cultivation. Tuve and Wilson exemplified the first tendency, whereas for the second, J. Robert Oppenheimer and his early education at the Ethical Culture School in New York City (founded upon the principles of Felix Adler) or Hans Bethe's inculcation with the ideals of *Bildung* come readily to mind (Schweber 2000: chapters 2–3). As the late S. S. Schweber observed, elite physicists in the middle decades of the twentieth century believed in science as an individual and a communal activity that embodied specific virtues—"honesty, tolerance, trust, truthfulness, and cooperation"—that upheld both the epistemic character of science as well as its moral qualities (Schweber 2000: 101). By the 1970s, if one believes the accounts of David Kaiser and of Steven Shapin, the worldview for science defined by these late nineteenth century settings had largely given way to a cultural acceptance of science as a job, and scientists as just as materially concerned as anyone else (Kaiser 2004; Shapin 2008). The shift grew from a variety of circumstances—the massive growth of the physics profession in the mid-twentieth century, post-World War II affluence, the constricted ideological environment of the cold war, the social and political upheaval of the Vietnam and Watergate eras and the distrust of authority that followed, and an increasingly conservative economic environment in recent decades that gauges the value of intellectual activity through measurable effects in dollars and cents, rather than by enrichment of an abstract human spirit.[12]

[11] For an introductory overview of the problem of science and nation, see Krige and Wang 2015.

[12] Paul Forman has also postulated a full-blown transition to an age of post-modernity, as opposed to simply a string of trends, in order to capture how, as he sees it, an entirely different set of values has come to predominate the broader culture since the 1970s (Forman 2007).

In one sense, one could say that the pure science ideal collapsed for good reason—it was not tenable in an era in which the political and institutional realities of scientific research militated so strongly against the basic tenets of science's governing ideology. At the same time, no better alternative has emerged for a public philosophy that can support basic scientific research and the authority of science as a form of public knowledge, not to mention fact itself as the basis for considered decision-making in a liberal democratic polity. In essence, the dilemma that Oppenheimer pinpointed in 1945 of the decoupling of epistemic from moral virtue has now come full circle in twenty-first century political life in the United States. Americans are beginning to discover the dangers of living without either epistemic or moral virtue in an age of "truthiness," the witty term that comedian and political commentator Stephen Colbert coined in 2006 to describe the growing political power of truths liberated from fact. Colbert presciently identified a burgeoning assault on facticity itself, a trend that threatens the entire modernist project of objective, incontrovertible fact as a basis for political economy, governance, and human action.[13] Recent discussions of the rise of a "post-truth" society in the aftermath of the 2016 U.S. Presidential election have further underscored the erosion of fact-based truth as the basis for the intelligent management of human affairs (Schuessler 2016; Sullivan 2016). After Hiroshima and Nagasaki, physicists and citizens had to wonder whether it was really "good to find out." Present-day politics, by contrast, exposes the heavy risks of not knowing, or of deliberately refusing to know.

Tellingly, in recent years some of the leaders of social constructivist analyses of science in the late 1970s and 1980s—most notably Bruno Latour—have called for a reconstruction of scientific authority as the only way to counter the corrosive effects of a political culture that dismisses scientific evidence as nothing more than an expression of special interests or socially constructed truths no more valid than any other truth claims (Latour 2013: 2–7). For Latour, global climate change makes such an effort an absolute imperative. What form that reconstruction might take, and whether it is even possible, I do not know, but it probably requires a willingness to embrace some kind of aspirational ideology for science and to temper an instinct to reduce scientific motivations to material interests, despite a post-1960s sensibility that immediately casts a suspicious eye upon overtures to transcendence, privileges the economic bottom line over appeals to abstract value claims, and increasingly dismisses fact itself as a mere reflection of individual bias. It is easy today to dismiss the pure science ideal and its vision of moral virtue as naïve or as a form of denial. It is harder to understand and admit what we have potentially lost by denying the legitimacy of earlier aspirations. A twenty-first century dismissal of old notions of self-discipline, moral rectitude, and the virtues of knowing may signify less a latter-day, worldly sophistication than a marker of our own contemporary surrender to the neoliberal veneration of the market, a calculating, instrumental view of the world, and the nihilistic individualism of truthiness. There is certainly a place for thinking

[13] On the historical rise of facticity in the context of English political economy from the sixteenth to the nineteenth centuries, see Poovey 1998.

ill of the past for its platitudes. But doing so also risks a lack of introspection and self-judgment towards our own times and our own failures of political and moral imagination.

**Acknowledgments** The author and the editors gratefully acknowledge the University of California Press for permission to reprint material that originally appeared in Jessica Wang, "Physics, Emotion, and the Scientific Self: Merle Tuve's Cold War," *Historical studies in the natural sciences* 42 (November 2012): 341–88.

## Bibliography

### *Primary Sources (Unpublished)*

Tuve, Merle A. 1947a. Notes for speech at the Westinghouse Sixth Annual Science Talent Search, 1 March 1947. Box 122, Folder Speeches Clippings etc., Merle A. Tuve papers. Manuscripts Division, Library of Congress, Washington, DC. Hereafter cited as Tuve papers.

———. 1947b. Letter to Paul C. T. Kwei, 3 Feb 1947. Box 123, Folder 1 Foreign Cooperation, Tuve papers.

———. 1950. Letter to William Webster, Chairman, Research and Development Board, 24 Mar 1950. Terrestrial Magnetism, Director 1950–1951, Carnegie Institution of Washington Administration Records, 1890–2001, Carnegie Institution of Washington, Washington, DC.

———. 1958. Handwritten lecture titled Conference discussion on Science and Secondary Education, St. Alban's School Washington Cathedral, 23 March 1958, Remarks by Dr. Merle A. Tuve. Box 179, Folder Tuve—Speeches & Conference, Tuve papers.

———. 1959. Letter to C. Gresham Marmion, 29 Sep 1959a. Box 196, Folder Bishop Marmion, Tuve papers.

———. n.d. [1959?] Handwritten draft headed Introduction—First Part. Box 186, Folder DTM Annual Report July 1959, Tuve papers.

———. n.d. [1960?] Handwritten draft headed Introd. Box 197, folder DTM Annual Report July 60, Tuve papers.

### *Primary Sources (Published)*

*FermiNews*. 1978. 21 Tons of Art. *FermiNews* 1 (22 June 1978). http://history.fnal.gov/art_architecture.html#21. Accessed June 2016.

Gibney, Elizabeth. 2014. China Plans Super Collider. *Nature* 511 (22 July 2014), online ed. http://www.nature.com/news/china-plans-supercollider-1.15603. Accessed June 2016.

National Research Council, Physics Survey Committee. 1966. *Physics: Survey and Outlook*. Washington, DC: National Academy of Sciences.

Oppenheimer, J. Robert. 1945. Speech to the Association of Los Alamos Scientists. Los Alamos, New Mexico, 2 November 1945. Reprinted on pp. 49–58 in Pais, Abraham. *J. Robert Oppenheimer: A Life*. Oxford: Oxford University Press.

Overbye, Dennis. 2006. China Pursues Major Role in Particle Physics. *New York Times*, 5 December 2006, online ed. http://www.nytimes.com/2006/12/05/science/05china.html?pagewanted=all. Accessed June 2016.

Rosen, Louis. 1971. Relevance of Particle Accelerators to National Goals. *Science* 173: 490–497.
Schuessler, Jennifer. 2016. 'Post-Truth' Defeats 'Alt-Right' as Oxford's Word of the Year. *New York Times*, 15 November 2016, online ed. http://www.nytimes.com/2016/11/16/arts/post-truth-defeats-alt-right-as-oxfords-word-of-the-year.html?_r=0. Accessed Dec 2016.
Sullivan, Margaret. 2016. The post-truth world of the Trump administration is scarier than you think. *Washington Post*, 4 December 2016, online ed. https://www.washingtonpost.com/lifestyle/style/the-post-truth-world-of-the-trump-administration-is-scarier-than-you-think/2016/12/02/ebda952ab897-11e6-b994-f45a208f7a73_story.html?hpid=hp_hp-top-table-main_sullivan-345pm%3Ahomepage%2Fstory&utm_term=.ceefcf1b6034. Accessed Dec 2016.
Weinberg, Alvin M. 1963. Criteria for Scientific Choice. *Minerva* 1: 159–171.
Weisskopf, Victor F., Chairman, High Energy Physics Advisory Panel. 1970. *Report on High Energy Physics 1969*, copy in U.S. Congress, Joint Committee on Atomic Energy, *AEC Authorizing Legislation, Fiscal Year 1971*, 91st Cong., 2nd sess., 3 and 5 March 1970, part 2, Appendix 2.
Wilson, Robert Rathbun. 1968. The Richtmyer Memorial Lecture: Particles, Accelerators, and Society. *American Journal of Physics* 36: 490–495.
———. 1970a. My Fight Against Team Research. *Daedalus* 99: 1076–1087.
———. 1970b. The Conscience of a Physicist. *Bulletin of the Atomic Scientists* 26: 30–34.
———. 1972. Testimony of Robert R. Wilson before the Joint Committee on Atomic Energy, Subcommittee on Research, Development, and Radiation, 29 February 1972. *AEC Authorizing Legislation Fiscal Year 1973*, 92nd Cong., 2nd sess., p. 1443.
———. 1977. From Frontiersman to Physicist. *Physics in Perspective* 2 (2000): 141–203. Adapted and published version of an interview by Spencer R. Weart, 19 May 1977.
———. 1987. Starting Fermilab: Some Personal Viewpoints of a Laboratory Director (1967–1978). *Annual Report of the Fermi National Accelerator Laboratory*. http://history.fnal.gov/GoldenBooks/gb_wilson2.html. Accessed June 2016.
Yuan, Luke C. L., ed. 1965. *Nature of Matter: Purposes of High Energy Physics*. Long Island Brookhaven National Laboratory, Associated Universities, Inc.

## *Secondary Literature*

Bourke, Joanna. 2003. Fear and Anxiety: Writing about Emotion in Modern History. *History Workshop Journal* 55: 111–133.
Costigliola, Frank. 2004. 'Like Animals or Worse': Narratives of Culture and Emotion by U.S. and British POWs and Airmen behind Soviet Lines, 1944–1945. *Diplomatic History* 28: 749–780.
Daston, Lorraine, and Peter Galison. 2007. *Objectivity*. New York: Zone Books.
Daston, Lorraine, and H. Otto Sibum. 2003. Introduction: Scientific Personae and Their Histories. *Science in Context* 16: 1–8.
Dennis, Michael Aaron. 1997. Historiography of Science: An American Perspective. In *Science in the Twentieth Century*, ed. John Krige and Dominique Pestre, 1–26. Amsterdam: Harwood Academic Publishers.
Ekbladh, David. 2010. *The Great American Mission: Modernization and the Construction of an American World Order*. Princeton: Princeton University Press.
Forman, Paul. 1987. Behind Quantum Electronics: National Security as Basis for Physical Research in the United States, 1940–1960. *Historical Studies in the Physical and Biological Sciences* 18: 149–229.
———. 2007. The Primacy of Science in Modernity, of Technology in Postmodernity, and of Ideology in the History of Technology. *History and Technology* 23: 1–152.
Gilbert, James. 1997. *Redeeming Culture: American Religion in an Age of Science*. Chicago: University of Chicago Press.

Herzig, Rebecca M. 2005. *Suffering for Science: Reason and Sacrifice in Modern America*. New Brunswick: Rutgers University Press.

Hilts, Philip J. 1984. *Scientific Temperaments: Three Lives in Contemporary Science*. New York: Simon and Schuster.

Hoddeson, Lillian, Adrienne W. Kolb, and Catherine Westfall. 2008. *Fermilab: Physics, the Frontier, and Megascience*. Chicago: University of Chicago Press.

Hughes, Thomas P. 1989. *American Genesis: A Century of Invention and Technological Enthusiasm, 1870–1970*. New York: Viking Press.

Jungk, Robert. 1969. *The Big Machine*. London: Andre Deutsch.

Kaiser, David. 2004. The Postwar Suburbanization of American Physics. *American Quarterly* 56: 851–888.

Kline, Ronald. 1995. Construing 'Technology' as 'Applied Science': Public Rhetoric of Scientists and Engineers in the United States, 1880–1945. *Isis* 86: 194–221.

Krige, John, and Jessica Wang. 2015. Introduction. *History and Technology*, special issue, *Nation, Knowledge, and Imagined Futures: Science, Technology, and Nation-Building, Post-1945* 31: 171–179.

Latour, Bruno. 2013. *An Inquiry into Modes of Existence: An Anthropology of the Moderns*. Cambridge, MA: Harvard University Press.

Lederman, Leon with Dick Teresi. 1993. *The God Particle: If the Universe is the Answer, What is the Question?* Boston: Houghton Mifflin.

Leslie, Stuart W. 1993. *The Cold War and American Science: The Military-Industrial-Academic Complex at MIT and Stanford*. New York: Columbia University Press.

McDougall, Walter A. 1985. *The Heavens and the Earth: A Political History of the Space Age*. New York: Basic Books.

McGrath, Patrick J. 2002. *Scientists, Business, and the State, 1890–1960*. Chapel Hill/London: University of North Carolina Press.

Needell, Allan. 1987. Lloyd Berkner, Merle Tuve, and the Federal Role in Radio Astronomy. *Osiris* 3: 261–288.

Poovey, Mary. 1998. *A History of the Modern Fact: Problems of Knowledge in the Sciences of Wealth and Society*. Chicago: University of Chicago Press.

Proton News, n.d. Robert R. Wilson: Remembered as 'Father of Proton Therapy' and Achievements in Physics and Medicine. *Proton News.*. www.proton-therapy.org/pr10.htm. Accessed June 2016.

Schrecker, Ellen W. 1986. *No Ivory Tower: McCarthyism and the Universities*. New York: Oxford University Press.

Schweber, Sylvan S. 1992. Big Science in Context: Cornell and MIT. In *Big Science: The Growth of Large-Scale Research*, ed. Peter Galison and Bruce Hevly, 149–183. Stanford: Stanford University Press.

———. 2000. *In the Shadow of the Bomb: Oppenheimer, Bethe, and the Moral Responsibility of the Scientist*. Princeton: Princeton University Press.

Shapin, Steven. 2008. *The Scientific Life: A Moral History of a Late Modern Vocation*. Chicago: University of Chicago Press.

Simpson, Christopher, ed. 1999. *Universities and Empire: Money and Politics in the Social Sciences during the Cold War*. New York: New Press.

Smith, Alice Kimball. 1965. *A Peril and a Hope: The Scientists' Movement in America, 1945–47*. Chicago/London: University of Chicago Press.

Smith, Michael. 1983. Selling the Moon: The U.S. Manned Space Program and the Triumph of Commodity Scientism. In *The Culture of Consumption: Critical Essays in American History, 1880–1980*, ed. Richard Wightman Fox and T.J. Jackson Lears, 177–209. New York: Pantheon Books.

Stearns, Peter N., with Carol Z. Stearns. 1985. Emotionology: Clarifying the History of Emotions and Emotional Standards. *American Historical Review* 90: 813–836.

Wang, Jessica. 2012. Physics, Emotion, and the Scientific Self: Merle Tuve's Cold War. *Historical Studies in the Natural Sciences* 42: 341–388.
Williams, Raymond. 1977. *Marxism and Literature*. Oxford: Oxford University Press.
Wickberg, Daniel. 2007. What Is the History of Sensibilities? On Cultural Histories, Old and New. *American Historical Review* 112: 661–684.

# Religious and Scientific Virtues: Maxwell, Eddington, and Persistence

**Matthew Stanley**

**Abstract** A critical but often overlooked epistemic virtue is persistence: when scientists come up against an obstacle, why should they proceed? An investigator needs to have some reassurance that further investigation will be fruitful despite setbacks, uncertainty, or confusing results. Two important physical scientists, James C. Maxwell and Arthur S. Eddington, drew on their religious beliefs and practices to help resolve this issue. They had very different religious identities, assumptions about the nature of scientific knowledge, and ways of experiencing the divine. Despite these divergences, Maxwell and Eddington constructed very similar epistemic virtues and related scientific practices.

**Keywords** Epistemic virtues • Persistence • Unity of nature • History of physics • History of astronomy • Science and religion • James Clerk Maxwell • Arthur Stanley Eddington

## Introduction

No experiment works the first time. No equation worth knowing presents its solution instantly. Sometimes the pieces simply do not fit. But even with such obstacles, we would be quite surprised if scientists immediately discarded a project when they ran into trouble. Indeed, we celebrate those scientists who pursue doggedly their goals even when discouraged. If Einstein had given up during the frustrating ten years between his special and general relativity theories, we would mark it as one of the great failures in the history of thought. This suggests a particular epistemic virtue, often overlooked: persistence. Why should scientists continue when they encounter difficulty? What do they do when things do not make sense? Why and how should they spend energy on a recalcitrant puzzle?

If we think of epistemic virtues as markers of good scholarly conduct, persistence is an important example. But as with many such markers, it might seem so obvious

M. Stanley (✉)
New York University, New York, NY, USA
e-mail: ms5100@nyu.edu



J. van Dongen, H. Paul (eds.), *Epistemic Virtues in the Sciences and the Humanities*, Boston Studies in the Philosophy and History of Science 321,
https://doi.org/10.1007/978-3-319-48893-6_4

that it requires no explanation. *Of course* good scholars persist. But persisting can actually be quite challenging to justify – if we find evidence contrary to our ideas, surely that means we should discard our plans rather than pursue them in the face of failure? Persistence implies that the investigator has a good reason to continue on their path even if indicators are otherwise. It implies an expectation that the world can, and should, be understood in a particular way. In a sense, recognizing persistence as a contingent epistemic virtue raises the question of why one should bother doing science at all – why should you expect to be able to understand the world?

Around 1900, the justifications for scientific persistence were in flux. The traditional justifications reaching back to the seventeenth century usually involved some form of religious reasoning - some conceptions of divine creation and action provided assurance that puzzle solving was a worthwhile activity. However, the term "religion" covers a very wide range of beliefs and practices, and it is important to be specific about the religious concepts involved and their connection to scientific virtues.

This essay will examine two cases in which we can see meaningful connections between religious beliefs and the virtue of persistence. James Clerk Maxwell (1831–1879) and Arthur Stanley Eddington (1882–1944) were both world-class scientists who were also deeply religious, though in quite different ways. They are representatives of two basic ways to approach the problem – a tight relationship between God and nature, and a loose relationship between God and nature. This roughly corresponds to conservative and liberal theology, historically. These two examples will demonstrate how religious virtues can be connected to scientific virtues, and hopefully shed some light on the general problem of how epistemic virtues form.

## Maxwell

Maxwell's name is best known today through the equations named after him. These were the first interrelated mathematical descriptions of the behavior of electricity, magnetism, and light (though he never wrote down the quartet that modern physics students memorize).[1] His extraordinary achievement was showing that electricity, magnetism, and light were all unified - that is, that what looked to be different phenomena and forces were actually just different manifestations of a deeper unity. He also helped shape thermodynamics, including providing the first demonstration that heat must be due to the motion of microscopic particles. In his spare time, invented the ophthalmoscope (by testing it on his dog), took the world's first color photograph, and wrote terrible poetry.

Maxwell was also an evangelical Christian. He was raised in both the Anglican and Presbyterian traditions, and as a young man in Cambridge came to evangelicalism through a powerful conversion experience. Evangelical Christianity was not a separate

[1] The most useful biography of Maxwell (despite some Victorian hagiography) remains Campbell and Garnett 1882. Most of Maxwell's manuscripts are collected in Harman 1990.

sect, but rather an ecumenical outlook that cut across denominations. That outlook was typically associated with a deep respect for scripture and an emphasis on a personal relationship with Jesus Christ, rather than institutional authority. Like most evangelicals, Maxwell thought of humanity as deeply sinful and in need of redemption from a wholly-other divine power. His God was the creator and lawgiver of Genesis who carefully crafted a universe for human beings.[2]

In considering the epistemic virtue of persistence, it is helpful to examine Maxwell's efforts to unify the laws of nature (that is, to reduce them to a single force, instead of multiple individual forces). By the time he finished college, there were already tantalizing indications that magnetism was related to electricity but it was not clear exactly how or why.[3] Many other investigators had made efforts to solve the puzzle, to little avail. Neither the laboratory nor theoretical evidence suggested any particularly fruitful routes of study. Maxwell, however, persevered. He developed a mechanical model of the ether in which electrical and magnetic effects could be explained by the rotation of vortex molecules and the movement of interspersed "idle wheel" particles. Electrostatic effects were accounted for by the elasticity of the medium made up of these molecules and particles. An unexpected effect of such elasticity was the possibility that the medium could support transverse waves of electromagnetic effects, which Maxwell realized would travel at the speed of light.[4] This suggested a profound conclusion, "that the luminiferous and the electromagnetic medium are one" (Maxwell 1861: 686).

Maxwell embraced the unification of natural laws suggested by his mechanical model. He was convinced that there was a true connection between optics and electromagnetism – that there was a fundamental principle hidden in the chaos of observable phenomena. For him, such unification was essential to the very notion of physical explanation: "When any physical phenomenon can be described as an example of a general principle which is applicable to other phenomena the phenomenon is said to be explained" (Maxwell 1874: 172).

Unity was a guideline for doing science – a goal - that worked very well for Maxwell and eventually resulted in his famous equations. He sought unity despite difficulty. What reason did he have for taking unity seriously as a guideline? Why did he think that nature was unified? Why did he think that magnetism and electricity would eventually fit together neatly? Why did he persist in searching for unity?

It was not that he had a naïve belief in the unity of nature. Maxwell was well aware of the psychological quirks of the human psyche and seriously considered the possibility that the unification of natural laws was only a feature of the mind and not the physical world. As a young man he wondered: "[A]re we to conclude that these various departments of nature in which analogous laws exist, have a real interdependence;

---

[2] Maxwell's religious views are discussed in Stanley 2015 and Theerman 1986. A useful overview of British evangelicalism is Bebbington 1989. On evangelicalism and intellectual life (particularly science) see Hilton 1991, Livingstone 1999, and Fyfe 2004.

[3] On the development of Maxwell's theory overall see Harman 1998, Siegel 1985, 1991, Smith 1998, Wise 1982, Buchwald 1985, and Schaffer 1995.

[4] See Harman 1998, 64–67, 102–108, and Smith 1998, 226–7.

or that their relation is only apparent and owing to the necessary conditions of human thought?" (Maxwell 1856a: 376–77).

Maxwell knew that the human mind sometimes sought out certain kinds of patterns even if those patterns were in reality absent. He thought it was entirely possible that the concept of an orderly, unified, well-organized universe was simply a human construction that we projected onto the world. To illustrate this danger, Maxwell presented two possible metaphors for the laws of nature:

> Perhaps the "book," as it has been called, of nature is regularly paged; if so, no doubt the introductory parts will explain those that follow, and the methods taught in the first chapters will be taken for granted and used as illustrations in the more advanced parts of the course; but if it is not a "book" at all, but a *magazine*, nothing is more foolish to suppose that one part can throw light on another (Maxwell 1856a: 381–2).

If nature was like a book, then there was a single unified argument. There was a common thread holding together the text that could be used to interpret and understand the whole even if you were only able to read one chapter. If so, then in physics, electricity could help you understand magnetism because they were both part of a single "document." In that case it was worth being persistent in looking for unity.

But if nature were like a magazine, where the separate articles had nothing to do with one another and could even be written by different authors, there was no such assurance. A magazine has no single argument. Rather, it is a collection whose elements may or may not have a connection to each other. There would be no guarantee that any one article could help the reader understand any other. There would be no reason to persist in looking for unity.

If this were the case, there would be no reason to think that electricity could help us understand magnetism. It was not obvious why scientists should choose one metaphor over the other. Whether nature was like a book or a magazine was of the highest importance for understanding it. But how was one to decide? Maxwell's solution was that he knew the book of nature's author – God – and he knew how that author wrote.

Maxwell thought of natural laws as being 'parts of one universal system.' He felt assured that there was a plan to the interrelationship of natural laws. This interrelationship was a way that God communicated His existence, and it was the unity of laws that revealed this communication. An 'arbitrary' distribution of individual laws (like the articles of a magazine) would not suggest anything about a divine plan, but unification (like the chapters of a book) would be highly improbable and therefore was a kind of divine communication. God had a plan for the world, and part of that plan was designing natural laws to fit together like the pieces of a puzzle.

Maxwell often described the importance of unity in his God's plans. In a letter to Bishop Charles John Ellicott he discussed the influence that religious thought should have on scientific investigators. He warned against trying to prove scripture via physics, instead focusing on what a man of science could draw from each realm:

> I think that each individual man should do all he can to impress his own mind with the extent, the order, and the unity of the universe, and should carry these ideas with him as he reads such passages as the 1st Chap. of the Ep. to Colossians. .., just as enlarged conceptions of the extent and unity of the world of life may be of service to us in reading Psalm viii.; Heb. ii. 6, etc. (Maxwell 1876: 418).

For Maxwell, a divinely unified universe was something that could be found in the Bible, not just the laboratory. Persisting in the scientific goal of a unity of nature was therefore encouraged by theology. The scriptural passages Maxwell referred to in this letter emphasized God's role as creator of the natural world ("For by him were all things created, that are in heaven, and that are in earth, visible and invisible") and the awe that God designed his creation for man ("What is man, that thou art mindful of him?") (Colossians 1:16, Hebrews 2:6, and Psalm 8, NIV). Thus Maxwell was powerfully linking the unity and order of nature not just with divine creation itself, but also with the role of humanity in that creation. In the same letter he argued that we can see "wisdom and power" in the uniformity of natural laws just as effectively as in the beneficial adaptations of living creatures: "uniformity, accuracy, symmetry, consistency, and continuity of plan are as important attributes as the contrivance of the special utility of each individual thing" (Maxwell 1876: 417). A properly evangelical reading of scripture could, he thought, justify his persistence.

It is important to note that Maxwell did not think that any particular scientific hypothesis proved, or was proven by, any specific religious statements. He was willing to make broad, general claims about the theology of nature (such as unity), but was quite hesitant to directly link particular scientific ideas to any part of religious doctrine:

> But I should be very sorry if an interpretation founded on a most conjectural scientific hypothesis were to get fastened to the text in Genesis, even if by so doing it got rid of the old statement of the commentators which has long ceased to be intelligible. The rate of change of scientific hypothesis is naturally much more rapid than that of Biblical interpretations, so that if an interpretation is founded on such an hypothesis, it may help to keep the hypothesis above ground long after it ought to be buried and forgotten (Maxwell 1876: 418).

That is, trying to fuse scripture and science was dangerous for both. Further, he warned that any ties between religion and science should be sought by individual Christians, and could only reflect their own personal beliefs and experiences. This overlap should be found in ways of thinking and acting – what we would call epistemic virtues.

These virtues were important for the way Maxwell thought God had designed the laws of nature. Natural laws were designed with the special feature that they were meant to be discovered: "[Once we understand some science] we are prepared to see in Nature not a mere assemblage of wonders to excite our curiosity but a systematic museum designed to introduce us step by step into the fundamental principles which are displayed in the works of Creation" (Maxwell 1856b: 71).

In particular, the unity of laws was intended for discovery. The connections of natural laws were 'systematic' in that they were carefully designed to attract the attention of humans and lead them to deeper and deeper principles. Laws were laid out like a trail of bread crumbs to guide the attentive person from diverse phenomena to unification via strategic connections. Scientific investigation, then, and particularly the search for unified physical laws, was a task given divine assent and even encouragement. Maxwell's God wanted him to understand the world in deeper and deeper terms – to push through the puzzles. His theology gave him a powerful set of tools for understanding the natural world, and for guiding his investigations in physics. He argued that God made the universe obey laws that were fundamentally unified and that He wanted humans to discover that unity:

> Is it not wonderful that man's reason should be made a judge over God's works, and should measure, and weigh, and calculate, and say at last "I understand I have discovered – It is right and true" … we see before us distinct physical truths to be discovered, and we are confident that these mysteries are an inheritance of knowledge, not revealed at once, lest we should become proud in knowledge, and despise patient inquiry, but so arranged that, as each new truth is unraveled it becomes a clear, well-established addition to science, quite free from the mystery which must still remain, to show that every atom of creation is *unfathomable* in its perfection (Maxwell 1856b: 77).

Discovering the design of these laws was not meant to be an easy victory, however. Their revelation was closely balanced against the deeper truths which humans could never know. God's intent was to encourage us to always be investigating further, not to be satisfied with what we already had:

> While we look down with awe into these unsearchable depths and treasure up with care what with our little line and plummet we can reach, we ought to admire the wisdom of Him who has arranged these mysteries that we find first that which we can *understand* at first and the rest in order so that it is possible for us to have an ever increasing stock of *known* truth concerning things whose nature is absolutely incomprehensible (Maxwell 1856b: 77).

Note the distinctly evangelical warnings against human pride and arrogance, and the evocative image of man's limited powers represented by 'our little line and plummet.' The deepest truths of nature were simply beyond our understanding, except where God allowed us to explore. As with the evangelical position on sin and redemption, our ability to know anything about the universe was just the result of God's grace in making those things known and knowable. Comprehension of nature was the result of God's free choice to set up the laws of nature such that they could be understood, not just a result of human efforts. God wanted us to learn.

Maxwell's explanations for the importance of unity show that he did not see it solely as a scientific goal. Instead, it also had profound religious significance. It was a religious virtue to know that God intentionally created a mysterious, but comprehensible universe. A virtuous person pushed forward into that mystery assured that there were answers, and they were findable. This gave Maxwell confidence in the related epistemic virtue of persistence: even if the world looks messy, you should persist in looking for unity. His approach to this problem was typical of conservative religious thought: it postulated a close relationship between the physical world and its creator, and rested on the idea that the Book of Nature could provide insight into God's thoughts.

## Eddington

A generation later, also in Cambridge, we can see the same epistemic virtue being justified in a very different way. A.S. Eddington (1882–1944), the Plumian Professor of Astronomy there, is best known now among the public as the man who provided the first observational evidence for Einstein's theory of relativity. That was, of course, extremely important, but his scientific legacy is actually more significant in astronomy. He was one of the first theorists to understand the inner workings of stars, and laid the foundation for a century of stellar astrophysics. In addition to his technical work, he was also one of the great science popularizers of his day.[5]

But he is of interest to us due to his deep and abiding religious belief. Specifically, he was a lifelong member of the Religious Society of Friends, better known as the Quakers. The Quakers are a Protestant sect dating back to the seventeenth century that emphasizes the presence of God within everyone, and a related embrace of mysticism, pacifism, and social activism. As was typical of liberal religious thought, they contended that personal religious experience was primary, with scripture and religious institutions of secondary importance.

As with Maxwell, we will examine how a particular religious virtue intersected with the epistemic virtue of persistence. The relevant idea is what the Quakers called "seeking." This refers to the Quaker virtue of constantly exploring, searching, and looking for new things in both the spiritual and temporal worlds. This has an anti-dogmatic stance as a crucial element: one should not try to find complete certainty because this leads to stagnation and a refusal to accept new ideas. Quakers associated this idea with a mystical outlook. Mysticism requires an ability to accept new knowledge at any time, unlike a scripturalist or fundamentalist stance which requires true knowledge to be incapable of revision or improvement. A seeker is a pragmatist, using whatever knowledge and tools are useful, instead of worrying about whether they are "really true" in an absolute sense.[6]

Eddington tried to live his life in line with the ideal of seeking, and this played an important role in his work in astrophysics. It allowed him to make progress on a particularly difficult area of theoretical astronomy. In 1916 he began investigating a problem that had proven intractable for generations of astronomers: what, exactly, *are* stars? And, how do they shine?

Previous attempts to solve this problem involved creating theoretical models of stellar structure based on physical principles that were well-understood and validated, such as Newtonian gravity and classical thermodynamics.[7] These attempts failed uniformly, and the resulting models looked and behaved nothing like stars. James Jeans, the distinguished mathematical physicist and Eddington's rival, described the difficulty this way:

> The mathematician of today who wishes to devote his skill to the service of astronomy finds himself in a particularly difficult position. Problems in which mathematical analysis can start from the basis of assured physical facts . . . form a class of strictly limited extent. The

[5]The only major biography of Eddington is Douglas 1956, though it has significant gaps. Stanley 2007a examines the religious and scientific aspects of his life. On Quakers and science more broadly see Cantor 2005.

[6]Good overviews of Quaker history include Barbour 1988; Brinton 1965; Kennedy 2001; Isichei 1970.

[7]Hufbauer 1981, 1991, 2006; DeVorkin 2000; Bishop 1978.

> class comprises, roughly speaking, those problems whose physical basis is the law of gravitation alone (Jeans 1909: 72–4).[8]

But it was quite clear by the twentieth century that gravitation was not enough to explain why stars were hot. It seemed that some critical element was missing from physics, and its lack made it impossible to understand stars. This was a deductive approach: start from completely valid premises and the conclusion is certain to be true, but without total certainty in one's premises, one has nothing.

Eddington took a different approach to the question of the source of stellar energy. The most obvious feature about stars is that they shine, he maintained, and therefore models must account somehow for this output of energy. Astronomers and physicists knew nothing about the detailed behavior of the energy source, so Eddington made a pragmatic, simplifying assumption. Assume that the energy is generated abstractly throughout the star in the simplest way possible:

> It is clear that we cannot arrive at much certainty with regard to the conditions in a star's interior. .. the weak link in the present investigation is that I have assumed without much justification that [energy production] is constant throughout a star. I have given some evidence that if it is variable the general character of the results would not be greatly altered; and, as a step toward the elucidation of the problem of stellar temperatures, I plead to be allowed provisionally one rather artificial assumption (Eddington 1919: 17).

He justified his uncertain foundations by appealing to the possibility of making progress on a difficult problem. And he was able to make progress: with slight adjustments of his provisional assumption, Eddington was able to reproduce many of the observable characteristics of stars without knowing any of the details of where the energy comes from.

Eddington's first breakthrough was the realization that his simplified, approximate models could actually reproduce the observed relations between the masses and luminosities of stars. This curve was an excellent fit to the data, and was essentially the first success in theorizing about the interiors of stars. Not everyone was impressed with this result, however. Jeans argued that the fit between calculation and observation was meaningless because it was not based on firm deductive foundations:

> But my dissent from Professor Eddington's work is not one of numbers, it is one of mathematical principle. I do not say that his curve is numerically inaccurate, but that it should not be in his diagram at all. One unjustified physical assumption has reduced him from an area to a curved line; two such assumptions would have reduced him to a point; and three such assumptions would have proved that stars cannot exist at all (Jeans 1925: 32).

Eddington's persistence past the limits of deductive methods was seen as a weakness by Jeans. Whether forward progress or firm foundations was more important was not obvious, and became the basis of the famously vigorous debates between the two scientists at the Royal Astronomical Society (Stanley 2007b).

---

[8] A short but useful biographical treatment of Jeans is Smith 1977. Milne 1952 is more comprehensive but somewhat polemical: Milne had vigorous disagreements with both Eddington and Jeans on various technical matters, and much of the book is Milne justifying his past positions.

Eddington continued on with his work despite his colleagues' criticism. He manipulated his new mathematical models and compared them to observations, and the differences allowed him to infer some of the basic behaviors of the source of stellar energy. At the time there were still multiple possibilities for the physical mechanism, with the most likely candidates being either the annihilation of oppositely-charged particles or "transmutation" (what we today call fusion). Eddington argued that the competing theories should be judged on how well they allowed further scientific investigation:

> The theory of annihilation of matter is more fertile in astronomical consequences than the other forms of the subatomic theory, and for this reason alone it seems worth while to follow it up in detail. We shall not be greatly concerned with *how* the annihilation is accomplished; but it may perhaps be well to have a scheme in mind (Eddington 1926: 306).

Even after he constructed a useful theory of stellar structure, he was not content to let it stand as a finished product. He presented it as something that needed to be challenged, pushed, and most of all *used*. According to him manipulating a theory at the edge of its applicability helped to not just solve the problems at hand, but also indicate where further investigation would be useful. He described his treatment of stellar structure thus:

> In this calculation we have pressed the theory to an extreme degree. Our object is not so much to assert the truth of the conclusions, as to use every opportunity of discovering by comparison with observation the directions in which our approximate treatment may be improved (Eddington 1919: 20).

He intriguingly said a theory should not be like a building (a permanent structure to be admired) but rather like an engine (something to move one forward) (Eddington 1920: 357).

He justified his methodology by pointing out that he had made great steps forward in the understanding of stars. His interest in seeking new, if tentative, knowledge, rather than restricting himself to what was certain, had allowed him to bypass some of the problems of stellar astrophysics. The uniqueness of his approach was emphasized by the objections his contemporaries brought against him. They argued that physics could only be based on things that were known to be absolutely true, like the laws of mechanics, and Eddington's basic assumptions had no such justification. But Eddington argued, successfully, that the results of expanding our understanding justified the uncertain beginning.

Eddington also said this pragmatic, exploratory approach to science explained how science can make progress at all when its results were constantly being overturned. This discarding of knowledge was not the tragedy it seems; rather, the tragedy occurred when someone thought they know everything necessary about a subject. If a scientist tried to reify a theory or an experiment, they would no longer be able to move any further in their investigation. Instead, they needed to accept that any result was temporary, and was true only so far as it allowed further exploration. This made the danger of disproof into a benefit – it forced a scientist to continually improve. This was the justification of persistence – if the puzzle piece does not fit, look for another place to use it.

The crux of the matter was that Eddington was arguing for an open-ended scientific process. Proof was not to be valued; it was the ability to know *more* that was important. *Stars and Atoms*, a popular book explaining his technical theory, took pains to explain exactly what this meant:

> It would be an exaggeration to claim that this limited success is a proof that we have reached the truth about the stellar interior. It is not a proof, but it is an encouragement to work farther along the line of thought we have been pursuing. The tangle is beginning to loosen. The more optimistic may assume it is now straightened out; the more cautious will make ready for the next knot. .. We have taken present day theories of physics and pressed them to their remotest conclusions. There is no dogmatic intention in this; it is the best means we have of testing them and revealing their weaknesses if any (Eddington 1927: 40–1).

In his inimitable style, Eddington illustrated the value of seeking over dogmatism with a reevaluation of one of the Greek classics. The story of Daedalus and Icarus was usually told to admonish those who push too far, but Eddington provided a novel perspective:

> In weighing their achievements, there is something to be said for Icarus. The classical authorities tell us that he was only "doing a stunt," but I prefer to think of him as the man who brought to light a serious constructional defect in the flying machines of his day. So, too, in Science. Cautious Daedalus will apply his theories where he feels confident they will safely go; but by his excess of caution their hidden weaknesses remain undiscovered. Icarus will strain his theories to the breaking-point till the weak points gape. For the mere adventure? Perhaps partly; that is human nature. But if he is destined not yet to reach the sun and solve finally the riddle of its constitution, we may at least hope to learn from his journey some hints to build a better machine (Eddington 1927: 41).

Difficulty in a scientific investigation was not to be feared, it was to be welcomed as a route to further understanding.

This essay has concentrated on Eddington's early work in theoretical astrophysics, as those writings were clearer and more explicit about the epistemic virtues at play. However, his theoretical projects late in life, notably his attempts to construct a unified field theory, also display many of these virtues. Over the course of about two decades Eddington developed a highly intricate and idiosyncratic system seeking to unify quantum mechanics and relativity, culminating in his book *Fundamental Theory*.[9] His continual exploration and modification of these ideas provide many interesting examples of the sort of theoretical persistence discussed here, though *Fundamental Theory* was different enough that other factors may have been involved as well. Persistence was particularly important for this unification work – it was widely disparaged by his colleagues and had little support. Wolfgang Pauli dismissed it as "complete nonsense," suggesting that it be considered as "romantic poetry, not as physics" (Van Dongen 2010: 119). Even without any broad acceptance of his approach, Eddington continued on the path that seemed most fruitful to him.

---

[9]The strongest attempt to understand Eddington's theory on its own terms remains Kilmister 1994. Kragh 2011 describes the development of the theory and places it in context with similar contemporary attempts. Stanley 2007a, chapters 5 and 6, discuss the epistemological roots of that work.

Eddington's persistence in his theoretical work was very similar to the Quaker virtues of seeking that he embraced in his religious practice. As a seeker, he considered fundamental certainty (such as inerrant scripture) to be far less important than maintaining a living, transforming faith and a direct experience of God. Eddington felt that Quakerism was fundamentally mystical, and that mysticism required an open attitude toward the world. He liked to cite one of the Queries (a series of short questions meant to stimulate prayer and thought at Quaker meetings), which read: "Are you loyal to the truth and do you keep your mind open to new light, from whatever quarter it may arise?" Eddington took much of his thinking about this from the prominent contemporary Quaker Rufus Jones, who described the pragmatism of Quaker mysticism thus: "Fortunately we do not need to understand vital processes and energies of life before we utilize them and start living by them" (Kenworthy 1984: 124). As one of Jones' practical mystics, Eddington felt that religious truth should not be put on a pedestal and kept separate, but it should instead inspire one, and change the way one lives life. Certainty was not to be sought after in either science or religion. In one of his best-selling books, Eddington wrote:

> We seek the truth; but if some voice told us that a few years more would see the end of our journey, that the clouds of uncertainty would be dispersed, and that we should perceive the whole truth about the physical universe, the tidings would be by no means joyful. In science as in religion the truth shines ahead as a beacon showing us the path; we do not ask to attain it; it is better far that we be permitted to seek. .. You will understand neither science nor religion unless seeking is placed in the forefront (Eddington 1929: 22–23).

So science seemed to share a virtue with religion – persist in searching for knowledge, even if there will be no end. A complicated puzzle was just one more step on an unending quest.

Eddington's approach to being a religious scientist was typical of early twentieth century liberal religion: avoiding talk about God as creator or directly interacting with the natural world, and instead emphasizing personal religious experience and how a religious outlook can work well with science. The practice of science was encouraged by the spiritual virtue of seeking for truth and a pragmatic approach to all experience.

## Conclusion

This essay has examined two versions of the epistemic virtue we have called "persistence." It is, essentially, the willingness to push forward with a scientific investigation despite roadblocks. Over a span of two lifetimes we saw how this virtue was conceived of and defended in rather different ways. For Maxwell, the maxim was "have faith." For Eddington, it was "there is always more."

Not everyone agreed with their embrace of this virtue. As we saw, James Jeans argued that stellar interiors were simply unknowable and thought astronomers should simply accept their ignorance. While most scientists would agree that persistence was a useful virtue, exactly how to deploy that virtue was not always clear, or agreed on. Epistemic virtues can be adhered to in varying strengths and in complicated ways. Maxwell and Eddington had very different foundational principles justifying persistence, but their preferred practices were quite similar. This suggests that there could be a wide range of cultural resources that could result in the same epistemic virtues. Perhaps this is an argument for a kind of epistemic pluralism: science can thrive in many different cultures.

## Bibliography

Barbour, Hugh, and J. William Frost. 1988. *The Quakers*. New York: Greenwood Press.

Bebbington, David W. 1989. *Evangelicalism in Modern Britain: A History from the 1730s to the 1980s*. London: Unwin Hyman.

Bishop, Jeanne. 1978. Golden era of Theoretical Physics: The Black Box of Stellar Energy. *Griffith Observer* 42: 3–17.

Brinton, Howard. 1965. *Friends for 300 Years*. Wallingford: Pendle Hill.

Buchwald, Jed. 1985. Modifying the continuum: methods of Maxwellian electrodynamics. In *Wranglers and Physicists: Studies on Cambridge Physics in the Nineteenth Century*, ed. P.M. Harman, 225–241. Manchester: University of Manchester Press.

Campbell, Lewis, and William Garnett. 1882. *The Life of James Clerk Maxwell*. London: Macmillan and Co..

Cantor, Geoffrey. 2005. *Quakers, Jews, and Science: Religious Responses to Modernity and the Sciences in Britain, 1650–1900*. Oxford: Oxford University Press.

Darrigol, Olivier. 2000. *Electrodynamics from Ampère to Einstein*. Oxford: Oxford University Press.

DeVorkin, David. 2000. *Henry Norris Russell: Dean of American Astronomers*. Princeton: Princeton University Press.

Douglas, Allie V. 1956. *The Life of Arthur Stanley Eddington*. London: Nelson.

Eddington, Arthur S. 1919. On the Pulsations of a Gaseous Star and the Problems of the Cepheid Variables Part I. *Monthly Notices of the Royal Astronomical Society* 79: 2–22.

———. 1920. The internal constitution of the stars. *Observatory* 557: 341–358.

———. 1926. *The Internal Constitution of the Stars*. Cambridge: Cambridge University Press.

———. 1927. *Stars and Atoms*. London: Oxford University Press.

———. 1929. *Science and the Unseen World*. New York: Macmillan Company.

Fyfe, Aileen. 2004. *Science and Salvation: Evangelical Popular Science Publishing in Victorian Britain*. Chicago: University of Chicago Press.

Harman, Peter M., ed. 1990. *Scientific Letters and Papers of James Clerk Maxwell*. Vol. 1–3. Cambridge: Cambridge University Press.

———. 1998. *The Natural Philosophy of James Clerk Maxwell*. Cambridge: Cambridge University Press.

Hilton, Boyd. 1991. *The Age of Atonement: the Influence of Evangelicalism on Social and Economic Thought*. Oxford: Oxford University Press.

Hufbauer, Karl. 1981. Astronomers take up the stellar-energy problem, 1917–1920. *Historical Studies in the Physical Sciences* 11: 277–303.

———. 1991. *Exploring the Sun: Solar Science since Galileo*. Baltimore: Johns Hopkins University Press.

———. 2006. Stellar structure and evolution, 1924–1939. *Journal for the History of Astronomy* 127: 203–227.

Isichei, Elizabeth. 1970. *Victorian Quakers*. Oxford: Oxford University Press.

Jeans, James. 1909. Review of R. Emden's *Gaskugeln*. *Astrophysical Journal* 30: 72–74.

———. 1925. Untitled. *Observatory* 609: 32.

Kennedy, Thomas. 2001. *British Quakerism, 1860–1920: The Transformation of a Religious Community*. Oxford: Oxford University Press.

Kenworthy, Leonard S., ed. 1984. *Living in the Light: Some Quaker Pioneers of the 20th century*. Kennett Square: Friends General Conference and Quaker Publications.

Kilmister, Clive. 1994. *Eddington's Search for a Fundamental Theory*. Cambridge: Cambridge University Press.

Kragh, Helge. 2011. *Higher Speculations: Grand Theories and Failed Revolutions in Physics and Cosmology*. Oxford: Oxford University Press.

Livingstone, David, D.G. Hart, and Mark A. Noll, eds. 1999. *Evangelicals and Science in Historical Perspective*. Oxford: Oxford University Press.

Maxwell, James C. 1856a. Analogies. In *Scientific Letters and Papers of James Clerk Maxwell*, ed. P.M. Harman, vol. 1, 376–377. Cambridge: Cambridge University Press.

———. 1856b. Inaugural Lecture at Aberdeen. In Reginald Victor Jones, James Clerk Maxwell at Aberdeen, 1856–60. *Notes and Records of the Royal Society of London* 28(1973): 71.

———. 1861. Maxwell to Faraday October 19 1861. In *Scientific Letters and Papers of James Clerk Maxwell*, vol. 1, ed. P.M. Harman, 686. Cambridge: Cambridge University Press.

———. 1874. On the (Physical) Dynamical Explanation of Electric Phenomena. In *Scientific Letters and Papers of James Clerk Maxwell*, ed. P.M. Harman, vol. 3, 172. Cambridge: Cambridge University Press.

———. 1876. Maxwell to Charles John Ellicott, Bishop of Gloucester and Bristol, 22 November 1876. In *Scientific Letters and Papers of James Clerk Maxwell*, ed. P.M. Harman, vol. 3. Cambridge: Cambridge University Press.

Milne, Edward A. 1952. *Sir James Jeans*. Cambridge: Cambridge University Press.

Schaffer, Simon. 1995. Accurate measurement is an English science. In *The Values of Precision*, ed. M. Norton Wise, 135–172. Princeton: Princeton University Press.

Siegel, Daniel. 1985. Mechanical image and reality in Maxwell's electromagnetic theory. In *Wranglers and Physicists: Studies on Cambridge Physics in the Nineteenth Century*, ed. P.M. Harman, 180–202. Manchester: University of Manchester Press.

———. 1991. *Innovation in Maxwell's Electromagnetic Theory*. Cambridge: Cambridge University Press.

Smith, Crosbie. 1998. *The Science of Energy: A Cultural History of Energy Physics in Victorian Britain*. Chicago: University of Chicago Press.

Smith, Robert W. 1977. Sir James Hopwood Jeans 1877-1946. *Journal of the British Astronomical Association* 88: 8–17.

Stanley, Matthew. 2007a. *Practical Mystic: Religion, Science, and A.S. Eddington*. Chicago: University of Chicago Press.

———. 2007b. So simple a thing as a star: the Eddington-Jeans debate over astrophysical phenomenology. *British Journal for the History of Science* 40: 53–82.

———. 2015. *Huxley's Church and Maxwell's Demon: From Theistic Science to Naturalistic Science*. Chicago: University of Chicago Press.

Theerman, Paul. 1986. James Clerk Maxwell and religion. *American Journal of Physics* 54: 312–317.

Van Dongen, Jeroen. 2010. *Einstein's Unification*. Cambridge: Cambridge University Press.

Wise, M. Norton. 1982. The Maxwell Literature and British Dynamical Theory. *Historical Studies in the Physical Sciences* 13: 175–205.

# The Epistemic Virtues of the Virtuous Theorist: On Albert Einstein and His Autobiography

**Jeroen van Dongen**

**Abstract** Albert Einstein's practice in physics and his philosophical positions gradually reoriented themselves from more empiricist towards rationalist viewpoints. This change accompanied his turn towards unified field theory and different presentations of himself, eventually leading to his highly programmatic *Autobiographical Notes* in 1949. Einstein enlisted his own history and professional stature to mold an ideal of a theoretical physicist who represented particular epistemic virtues and moral qualities. These in turn reflected the theoretical ideas of his strongly mathematical unification program and professed Spinozist beliefs.

**Keywords** Epistemic virtues • Unified field theory • unification • Albert Einstein • Einstein's autobiography • Baruch Spinoza • Einstein and Spinoza • Einstein and religion • Einstein's philosophy • Philosophy of science

## Introduction

In the early nineteenth century, the experimental physicist Michael Faraday perfected his notebook recordkeeping such that the data would enter them entirely without regard to what his original hypotheses might have been. His way of working sharply contrasts with Arthur Worthington's, who in the earlier years of his career saw the need to generalize and brush over asymmetries in his visual studies of liquid drops: such asymmetries were deemed irrelevant for capturing the latter's essence. Faraday and Worthington represent two ideal types, two 'personae' that figured prominently in the practice of nineteenth century science. The novel and self-denying scientist aspiring to the objective representation of nature represented the world differently than the intuitively working scholar, who wished to point out the true essence of a natural phenomenon: the first ideally presented his observations

J. van Dongen (✉)
ITFA, Faculty of Science, University of Amsterdam, Amsterdam, The Netherlands
e-mail: J.A.E.F.vanDongen@uva.nl

 
J. van Dongen, H. Paul (eds.), *Epistemic Virtues in the Sciences and the Humanities*, Boston Studies in the Philosophy and History of Science 321,
https://doi.org/10.1007/978-3-319-48893-6_5

unfiltered and directly, without intervention; the second would at times, e.g., see the need to smooth out the irregular or asymmetric.

This is of course one of the key observations of Lorraine Daston and Peter Galison's study *Objectivity* (2007, 243–246). Clearly, the dual roles of the hypothesizing and observing scientist changed due to the rise of the epistemic virtue of objectivity in the course of the nineteenth century. The examples of Faraday and Worthington may be taken to also illustrate another key aspect of the role of epistemic virtues, however: they document how the nature and content of the *knowledge produced* changed as a different epistemic virtue, embodied by a different scholarly persona, rose to prominence.

This connection between knowledge and the roles of epistemic virtues and personae[1] is often overlooked. Nevertheless, it may reveal us hitherto unseen relations in knowledge production. To illustrate the point, we will study the example of Albert Einstein, who used his own biography and related ideal versions of the 'theoretical physicist' to argue for the virtuousness of his 'unified field theory' physics.

What is a 'scientific persona' and how may this concept relate to Einstein and his autobiography? Literature on 'personae' places the concept between individual biography and social institution: 'personae' (for example, the 'scientist' vs. the 'natural philosopher' vs. the 'poet') mark a cultural identity that is shared and serves as example among a collective body of scholars (Daston and Sibum 2003). A particular persona is reflective of a set of values, attitudes and scientific practices; it is both negotiated, internalized and projected among the practitioners of a discipline. It represents considerable agency and influence as disciplinary icon. In Einstein's time, theoretical physics was a fairly new sub-discipline,[2] coming into its own on the heels of the relativity successes: the persona of the 'theoretical physicist' was being molded while that of a 'physicist' *per se* was being reshaped. As the resistance to relativity exemplifies (see, e.g., Wazeck 2009; van Dongen 2010b), Einstein's science, biography and public stature were central in the realignment of what an ideal physicist and 'theorist' should be: abstractions of Einstein's person were projected on these personae while they were being (re-)created.[3]

As Einstein cared deeply about the future direction of physics and the plight of his later endeavours, he inadvertently but deliberately engaged his own history to argue for particular scientific choices and certain notions of what a proper theorist was and did. In various autobiographical accounts, and particularly in his 1949 *Autobiographical Notes* Einstein presented certain choices for theoretical virtues and methods as linked to certain epistemically virtuous attitudes. Furthermore, these choices were being promoted as representing both the epistemically and the *morally* right choice. On this point, one is reminded of Thomas Kuhn's essay on theory choice: in Kuhn's analysis, the weighing of various virtues such as 'simplicity' or 'accuracy' in theory choice is just *like* the weighing of values in moral

[1] See in this regard also Tai and van Dongen 2016; Tai 2017.

[2] For the sub-discipline's history, see Jungnickel and McCormmach 1986.

[3] See for example how young Werner Heisenberg related to Einstein and the latter's recollections in the mid-1920s: Heisenberg [1974] 1989, in particular 113–114.

judgment (Kuhn 1977). The example of Einstein will suggest that the relationship between these kinds of judgments need not only be seen as that of a mere analogy, but as at times direct and concrete.

This paper will then do two things: it will first show how indeed the promotion of particular epistemic virtues can be linked with concrete choices made in the practice of science, and thus how they affect its result: the choice of theory. We will study this relation in Einstein's motivations for his unified field theories. Secondly, we will see in Einstein's example how the distinctions between the moral and the cognitive may dissolve in the role of these virtues, and how in his case they were linked to the conviction that 'true' scientific pursuit is preconditioned on a religious gaze.

Strikingly, the relation between theory choice, theoretical virtues and epistemic virtues is laid bare when they meet in the ideal Einstein portrayed of himself later in life. What virtues can we identify and what was their role in his practice of science? Given Einstein's actual history and agenda, how can we understand the autobiographical accounts that he produced? Clearly, addressing these questions will not only help in understanding the relation between knowledge and the persons making it, but also aid in forming a comprehensive biographical and literary coherent understanding of Einstein. We begin by looking at Einstein's turn to unified field theory and how this fits into his broader development as a scholar.[4]

## Einstein's *Werdegang* from Empiricist to Rationalist

In 1905, his 'miracle year', Einstein famously proposed the first version of the theory of relativity, as it would later be known. It was essentially a reformulation of the electrodynamics of Hendrik Antoon Lorentz, but a reformulation in which experience stood at the epistemic core. Central was a new interpretation of the time parameter, which was motivated by a Machian sense-datum type intuition: "we should be aware that all our pronouncements in which time plays a role are always pronouncements about simultaneous events. For example, if I say that 'that train arrived here at 7 o'clock', I mean to say, more or less, that the arrival of the train and the pointing of the small hand of my clock at 7 are simultaneous affairs" (Einstein 1905, 892–893). Clearly, the author of this passage stands directly in the world of experience and looks around himself, from whence he constructs his concepts.

Later in life, in 1954, Einstein opined in a letter to a colleague, Louis de Broglie, that he had "long been convinced that one shall not be able to find [the right description of the quantum] in a constructive way from the known empirical relations between physical things, because the required mental leap would exceed human powers."[5] In this statement, experience and observation are put at a distance from the creative scientific process. The move reflects Einstein's then decades long involvement with unified field theory research. These theories—Einstein's attempts

[4] For a more extensive account of this aspect, see van Dongen 2010a.

[5] Einstein to Louis de Broglie, 8 February 1954, as quoted in van Dongen 2010a, 2.

at an alternative theory for the quantum—were foremost creative efforts based on mathematics and hardly made any connection to experience.

Einstein's early work, however, often offered direct experimental hypotheses, as in the case of his light quantum proposal of 1905. Revealingly, in 1925 Einstein wrote to his friend and fellow theoretical physicist Paul Ehrenfest on that very same subject that he "no longer think[s] about experiments on the boundary of waves and particles" as "inductive means will never get you to a sensible theory."[6] Einstein's change in attitude regarding the epistemic role of experience is also reflected in the difference in appreciation he awarded 'facts' (i.e. experientially validated facts about the world) in the recollection of his own struggle to attain his greatest achievement, the formulation of the general theory of relativity, completed in 1915. In 1918, he held that that struggle had taught him that trustworthy theories needed "to be built on generalizable facts."[7] In the 1949 autobiography, however, Einstein expressed that general relativity had taught him something entirely different: that there was "no way" to formulate a successful theory from a "collection of facts" (Einstein [1949], 89). Experience could only play a role at the end of the creative process, to check whether its product was valid or not.

When Einstein changed his recollections, he presented different histories of himself. This was not without programmatic intent, however. The 1918 statement on the pertinence of 'facts' was written in a personal letter to a friend, Michele Besso, but the 1949 recollection was offered quite publically: it was included in the *Autobiographical Notes* that were part of the highly visible volume "Albert Einstein: Philosopher-Scientist", dedicated to Einstein's science and philosophy. These *Notes* appear catered to promote Einstein's then long-standing research program in field theory unification for a well-chosen audience. The volume further contained a number of essays on physics and philosophy by prominent contemporaries to which Einstein wrote detailed replies. Its expected stature may be illustrated by the alternative two titles that the book's editor, Paul A. Schilpp, had contemplated: "(1) 'The Scientific Battle of The Twentieth Century,' and (2) 'The Future of Physics'" (Schilpp [1949], xvi). Clearly, Einstein, his co-contributors and their prospective readers were already well aware of his own status as opinion leader and iconic exemplar—in other words, of the prominent role that Einstein's biography played in the shared sense of what it meant to be a 'physicist', and a 'theoretical physicist' in particular.

The history of the young, experientially directed Einstein had become a familiar trope in epistemological discussions by the late 1920s. The logical empiricists, for example, had modeled their position that theories can be neatly divided in empirical propositions and coordinative definitions on Einstein's treatment of observation in relativity (Howard 1994). Likewise, experimental physicist Robert A. Millikan venerated Einstein on his seventieth birthday for having made "modern science essentially empirical", and he singled out the 1905 version of relativity which, according to Millikan, was supposedly created out of the notorious Michelson-Morley experi-

[6] Einstein to Paul Ehrenfest, 16 September 1925, as quoted in van Dongen 2007, 117.

[7] Einstein to Michele Besso, 28 August 1918 (Speziali 1979, 81).

ment on the (non-)existence of the ether (Millikan 1949). The older Einstein, however, actually distanced himself from logical empiricist philosophy and denied any direct role of the Michelson-Morley experiment in his creative process.[8] There are good reasons to doubt the older Einstein's recollections: for one, he had no problem attributing the Michelson-Morley experiment a prominent role in his earlier accounts of the events of 1905 (van Dongen 2009). Furthermore, the formulation of general relativity in 1915 certainly did depend on 'facts'.

Einstein's downplaying of the creative merit of experience went hand in hand with an increased emphasis on trusting "faith in [...] logical simplicity" of the laws as guide.[9] This meant that researchers should aim for the mathematical unification of theories: they should aspire to bring together under one all-encompassing mathematical construction all of the natural forces (at the time, that meant the forces of gravity and electromagnetism), together with space, time and matter. As Einstein's research moved in the direction of this project, he consciously or unconsciously adjusted his recollections. By all appearances, he did so, at least in part, in order to justify his choices and to influence his field.

The theory of quantum mechanics had been created with much direct input from experiment, unlike Einstein's efforts in unified field theory. He famously criticized quantum mechanics, initially for being inconsistent, and later for its alleged 'incompleteness' when considered from a realist perspective. Usually this aspect and the theories' probabilistic nature are cited as indications of a principally philosophical disagreement with the quantum theory. Yet, there is a complementary perspective which provides a more comprehensive view of Einstein's position: when assessing his positions on the quantum theory, one should also consider his positive program in field theory and indeed his ideas on how to properly do theoretical physics. The incompleteness of statistical descriptions was to be overcome by finding the proper equations for the "total field", i.e., logically simple and unified field equations and their ideal particle solutions (Einstein [1949], 81). As Einstein put it, the proper laws can foremost be found by "looking for the mathematically simplest concepts and the link between them. [...] The creative principle resides in mathematics"; there "lies the theorist's hope of grasping the real in all its depth" (Einstein [1933], 300–301). Another way that Einstein expressed the same maxim was to state that one should strive after formulations of the laws that displayed "inner perfection" or "naturalness" (Einstein [1949], 23); in the end, only familiar classical field theories could possibly live up to his intuitions regarding 'naturalness'. These kinds of theories also safeguarded his own sense of physical causality and locality. In this sense, then, his familiar criticisms regarding the problematic nature of quantum mechanics should also be seen as an advertisement for his own efforts in field theory (see e.g. Einstein 1936).

[8] Holton 1968, 1969; for a nuanced treatment of Einstein's relation to realism and empiricism, see also Howard 1993.

[9] Einstein to David Bohm, 24 November 1954, as quoted in van Dongen 2010a, 181–182; for more on Einstein and 'simplicity', see also Howard 1998.

Einstein judged quantum mechanics to be not only philosophically questionable but also mathematically unnatural. Werner Heisenberg's matrix formulation of the quantum theory, for example, was initially qualified as "a true witches' multiplication table."[10] Theory criticism was accompanied by implicit value judgments: the *Autobiographical Notes* assure us that "the faith" that experiential "facts by themselves can and should yield scientific knowledge" is based on a "prejudice" (Einstein 1949, 49). On another occasion, Einstein qualified the motivation for changing theories because of a conflict with experience "trivial, imposed from without"; unification and simplification of premises provided a "more subtle motive" (Einstein 1950, 13, 16). If the unification program fails, he confided to a young mathematician, Frenchman André Lichnerowicz, then we can only hope to understand things "superficially."[11] Einstein wrote another colleague that mere "humans are usually deaf to the strongest arguments, while they are constantly inclined to overestimate the accuracy of measurement."[12] The message is clear: the theorist who occupies himself with unified field theory is *virtuous*, as he wants more than superficial knowledge: Einstein insisted that a "true theorist" believes that comprehension is "built on premises of great simplicity" (Einstein 1950, 13).

Where did Einstein's emphasis on these virtues come from, and when exactly did they become so very prominent? And when did he begin altering his recollections of the road to relativity? This is really a chicken and egg problem, but one moment in Einstein's biography does stand out as of great importance for the change in his epistemological outlook: finding the final field equations of the general theory of relativity in November of 1915. In his search for 'generally covariant' and relativistic field equations for gravity, Einstein had balanced a mathematical and physical approach, which initially led to contradictions. In 1913, he preferred the answers that his physical demands were giving him and he in fact more or less gave up on the mathematical idea of general covariance, despite that this went against one of his initial key intuitions. Eventually he returned to his more balanced approach and the final theory was again generally covariant. Einstein first concluded that he had too easily given up on his mathematics, but in fact he had needed both physics and mathematics (Norton 1984, Janssen and Renn 2007; Renn and Sauer 2007).

As we already saw, he did not immediately change his epistemological outlook: those changes would grow on him. Much later, for instance in a prominent 1933 lecture on methodology in Oxford, his 1949 autobiography or in the 1954 letter to de Broglie, the experience was brought forward to bulwark a strongly worded and a much more one-sided methodological conviction:

> The gravitational equations could *only* be found by a purely formal principle (general covariance), that is, by trusting in the largest imaginable logical simplicity of the natural laws. [...] This is how I became a fanatic believer in the method of 'logical simplicity'.[13]

---

[10] Einstein to Michele Besso, 25 December 1925 (Speziali 1979, 215–216).

[11] Einstein to André Lichnerowicz, January 1954, as quoted in van Dongen 2010a, 5.

[12] Einstein to Max Born, 12 May 1952 (Born et al. 1971, 188–189).

[13] Einstein to Louis de Broglie, 8 February 1954, as quoted in van Dongen 2010a, 2–3; emphasis as in original.

The methodological recipe of unified field theories resonated with, and partly originated in the experience of general relativity.

It should be pointed out that striving for unity and simplicity was also present in Einstein's earlier work, and it is not hard to identify origins of these ideas in even his earliest youth: Einstein had been an avid reader of the popular science books of Ludwig Büchner and Aaron Bernstein and these emphasized the banishment of dualisms and a search for unity in concepts (Gregory 2000). Striving for a unified *Weltbild* was not unique to Einstein, as many of his generation (and many more since then) shared this desire: regardless of whether they had as avidly enjoyed Bernstein and Büchner or not, Einstein's German contemporaries shared a cultural heritage that included Fichte, Goethe's Faust and the philosophies of Von Humboldt and Schelling, who all emphasized a search for a unified world picture (Holton 1998). In an article eulogizing a close colleague, Max Planck, Einstein even called this search the scientist's "supreme task" (Einstein 1918, 246). What makes the later Einstein stand out from his generation and the generation that immediately followed him, however, is his increased emphasis on the near-exclusive creative merit of mathematics and his increased distancing from experience in favor of theory.

How were Einstein's ideas about the ideal theory and ideal theorist put to use concretely? His field theory unifications spanned a period of at least three decades. They included such abstract attempts as five dimensional generalizations of relativity in the hope to accommodate the electromagnetic force alongside the gravitational force (now known as 'Kaluza-Klein' theories; see van Dongen 2002, 2010a), or efforts to attain the same by considering space-time universes that were torqued rather than relativistically curved; these theories were called 'teleparallel' (Sauer 2006). All these theories were to reproduce the intrinsically discrete nature of the world captured by quantum mechanics and the integer valuedness of the electron charge. Presumably, their equations would allow solutions that avoided the problematic infinities typical for the general theory of relativity (known as 'singularities'). Einstein hoped that such solutions could model elementary particles in a natural fashion. He celebrated newly proposed equations invariably as of "great logical simplicity",[14] but abandoned them again when they did not produce the desired particle solutions. The virtue of 'simplicity' (or, in Einstein's usage, its near synonyms 'unity' and 'naturalness') guided these efforts, as the following words on his attempts at a mathematical unification of electrons and protons through so-called 'semivectors' testify:

> I discovered [...] a new sort of field, which we called 'semivectors'. These semivectors, after vectors, are the simplest mathematical fields that are possible in a four dimensional continuum, and they describe, in a natural way, essential properties of electrical particles. [...] The important point is that all these constructions, and the laws connecting them, can be arrived at by the principle of looking for the mathematically simplest concepts and the link between them (Einstein 1933, 301).

Unfortunately for Einstein, the semivector result alluded to here proved to be spurious as well, and he quickly dropped the concept (van Dongen 2004, 2010a).

[14] Einstein to Maurice Solovine, 23 December 1938 (Einstein 1956, 76).

## Unified Field Theory and the Morally Virtuous Theorist

Recounted recollection is a way to fashion our own self-image and an image of ourselves to the outside world. As we saw, in his autobiography and elsewhere, Einstein presented an edited version of his own history. In these accounts theoretical virtues and methodological precepts were reflected in the choices that an epistemically virtuous theorist was supposed to make. Einstein's example can show us more, however. We will argue that it makes clear that his ideal theoretical physicist was at the same time both morally and epistemically to be emulated: the virtuous had at the same time a moral *and* an epistemic charge, while epistemology carried religious connotations for Einstein.

Einstein considered choosing a more complicated mathematical structure over a simpler alternative to be 'sinful': "sin remains sin, even if it is committed by otherwise ever so respectable men" (Einstein 1949, 77). Objectionably complex were those theories that lacked a sufficient degree of unification. As we already saw, Einstein held up his senior Berlin colleague Max Planck as another example of a virtuous theoretical physicist. Planck had famously introduced the energy quantum in 1900 and Einstein wrote a eulogy for his sixtieth birthday in 1918. He expressed that Planck exhibited "inexhaustible patience and perseverance" while he "devoted himself [...] to the most general problems of our science, refusing to let himself be diverted to more grateful and more easily attained ends." In Einstein's eyes, Planck was a unifier, too, who might yet "succeed in uniting quantum theory with electrodynamics and mechanics in a single logical system" (Einstein 1918, 247–248). In the 1918 Planck eulogy, Einstein still awarded experience a creative role in "uniquely determin[ing] the theoretical system"; in his 1949 autobiography, however, Einstein claimed that Planck's early work on the quantum was "all the more remarkable because, at least in its first phase, it was not in any way influenced by any surprising discoveries of experimental nature" (Einstein 1949, 37). As anyone familiar with the history of the black body radiation problem (Kuhn 1978, Gearheart 2002) will agree, this is simply wrong, as Einstein likely would have known too. His words seem intended to substantiate a controversial characteristic of the unified field theory methodology with another example, aside from his own, rather than an attempt at rendering Planck's efforts accurately. Indeed, Einstein also produced histories of Johannes Kepler and James Clerk Maxwell that conveyed similar messages (Einstein 1930a, 1931).

The 1918 Planck eulogy stands out for another remarkable characteristic. Its evocative language and imagery, which addressed the practice of science in an at times exalted fashion, is easily compared to religiously inspired moral text. In the "temple of science", as Einstein chose his words, only true scientists would remain if "an angel of the Lord" were to drive out all the Philistines. Those who remain might fulfill that "supreme task [...] to arrive at those universal elementary laws from which the cosmos can be built up by pure deduction" (Einstein 1918, 244, 247). Here, religious metaphor emphasized the virtuous nature required to practice science properly: only those who aim at the universal—at "unity in the foundations"

(Einstein 1936)—are truly virtuous. Belief in such methodology bordered on religious belief: in 1950, Einstein wrote that the idea "that the totality of all sensory experience can be 'comprehended' on the basis of a conceptual system built on premises of great simplicity" was indeed "a miracle creed".[15]

Einstein also used exalted language in the Autobiographical Notes: he spoke of a "paradise" beyond the "merely-personal", the "contemplation" of which "beckoned like a liberation" in his early youth (Einstein 1949, 5). According to Lorraine Daston (2008), such abnegations of the personal are to be seen as an expression of the rise of objectivity during the same period. We should be aware, however, that to Einstein, the romantic genius is more of an ideal type than the self-denying record keeper. Einstein's paradise beyond the personal is not found in unfiltered maps of data points and immediate representations of observations, but in the unified universes captured by grand and speculative thought. As he wrote in his article on *Religion and science*: "individual existence [is] a sort of prison and one wants to experience the universe as a single significant whole." Not the nitty-gritty of experimental detail, but the leap of abstract and intuitive creativity motivates and liberates. Indeed, those "geniuses" that may attain such elevation possess a "cosmic religious feeling" (Einstein [1930b], 41). In this light, Einstein's scientific flight from the confining personal and "primitive feelings" (Einstein 1949, 5) was intended to move him beyond the partial and particular of incomplete descriptions: he wished to attain the impressive breadth and deductive certainty of all-encompassing and fully unified theories.

In the history of objectivity, Arthur Schopenhauer figures prominently. Daston and Galison believe that his instruction to suppress the 'Will', as a precondition to enable introspective reflection towards the apprehension of the principles of the world, is related to the objective scientist's self-restraint—his philosophy was a contemporary expression of the same sentiment (Daston and Galison 2007, 203–204). The objective persona, Daston (in a separate publication) further finds, eschewed the personal while immersed in solitudinous thought—yet, at the same time, he was part of a timeless community of scholars, who were imagined as equally "thinking selves dedicated to fathoming nature" (Daston 2008, 17). These scholars, then, would produce philosopher Thomas Nagel's 'view from nowhere': the same timeless and invariant facts and structures were to be unearthed by them, independent of any personal dimensions. In Daston's account, here is where Einstein fits the story of objectivity: his "paradise beyond the personal" could be attained by aiming at invariance and symmetry, which would be the same to all but only revealed to a few.

Einstein's unified field theory agenda and refashioned ideal theorist do resonate with this aspect of the history of objectivity: both are equally aimed at the universal. Indeed, Einstein approvingly identified Schopenhauer as someone who had understood that the "cosmic religious feeling" originates in the desire to "experience the universe as a single significant whole": in the "sublimity and marvelous order which reveal themselves in nature and in the world of thought" does one, à la Schopenhauer,

---

[15] As in Einstein 1950, 13; on this article, see particularly also Sauer 2014.

find a way to escape "the futility of human desires" (Einstein [1930b], 41). Einstein believed, "with Schopenhauer", that the ideal scholar, "a finely tempered nature", should wish to "escape from personal life into the world of objective perception and thought." That perception, however, is not the unmediated communication of direct observation, but the contemplative view on the impersonal and logically simple, "the restful contours apparently built for eternity" (Einstein [1918], 245, 251).

Another philosopher, at least as much read and revered by Einstein, captured a similar sentiment, even if he wrote and thought in an entirely different era: Baruch Spinoza. In the Weimar years Spinoza's philosophy went through a remarkable revival among German Jews as he was widely celebrated as a Jewish hero. His biography offered Weimar Jews the possibility to celebrate increasingly embattled liberal values and to still believe in assimilation, while affirming their ethnic authenticity as was particular to the Jewish renaissance of the Weimar years (Wertheim 2011). Einstein had read Spinoza throughout his life, beginning right after finishing his college education together with his friends of the mock 'Akademie Olympia', and on many occasions thereafter, even visiting Spinoza's small house in Rijnsburg near Leiden in 1920 (Stachel et al. 1989; Buchwald et al. 2006, 604); historian of science Max Jammer has actually identified Spinoza as Einstein's "favorite philosopher" (Jammer 1999, 144).

In Spinoza, rational contemplation of the world, moral virtue and a unified understanding of God and nature meet: Einstein's approval of his philosophy is an expression of what it meant to him to be both a virtuous scientist and a morally just person. Epistemic and moral virtue overlap here in a most direct manner. Spinoza, in his *Ethics*, argued (with geometric precision, that is, in a logical style modeled on Euclid's geometry) that man's moral well-being lies not in a life spent on pursuing the lower passions or the contemplation of the superstitions known as religion, but in a life spent on the rational pursuit of knowledge, particularly knowledge of the eternal laws of nature. Nature, in Spinoza's view, a view which Einstein warmly agreed with, was synonymous to an impersonal God whose existence was infinite and deterministic. According to Spinoza, our virtue, happiness and even salvation lie in understanding the universe: the more we know of the true ideas of God, the freer we are, since we will deal with life's disturbances with greater equanimity (Nadler 2001). Spinoza, too, saw in rational contemplation of unified eternity a flight from the "merely personal." Einstein agreed with Spinoza's positions, identified with Spinoza as a fellow Jew, and most famously followed him in his understanding of God: "I believe in Spinoza's God who reveals himself in the orderly harmony of what exists, not in a God who concerns himself with fates and actions of human beings."[16]

Einstein held that our moral and esthetic judgment are "tributary forms in helping the reasoning faculty toward its highest achievements"—one is right in "speaking of the moral foundations of science", even if the reverse, to reduce ethics to formulae, "must fail."[17] Science cannot give us a moral road map, but it provides a

[16] Einstein to Rabbi Herbert S. Goldstein, April 1929, on p. 49 in Jammer 1999.

[17] Einstein, 1930, interview cited in Jammer 1999, p. 69.

way to elevate man: "one does people the best service by giving them some elevating work to do and thus indirectly elevating them", Einstein wrote in a essayistic note entitled "Good and evil." He specified that such work could be in the arts or sciences, and in the latter case added that "to be sure, it is not the fruits of scientific research that elevate a man and enrich his nature, but the urge to understand" (Einstein [1934], 13). This reminds us of both Spinoza's *Ethics* and Einstein's unified field theory efforts—not just because Einstein asked us to leave out of consideration whether one's contemplations bear fruit, but primarily because of the insistence on "understanding." For, what did he believe "understanding" entailed? In Einstein's physics, to understand meant to have a 'simple' and unified description: "[We mean] by simplicity [...], that the system contains as few independent assumptions or axioms as possible; for the totality of logically independent axioms stands for the 'un-understood' remainder." In the same 1931 essay, he further made clear that his own efforts "first and foremost concern the logical unity in physics." Indeed, because "one wants to *understand* the existing, real world."[18] So Einstein's ideal theorist combines both epistemic and moral virtue in his attempt to unify: unified field theory science is virtuous science à la Spinoza, as it aims to elevate by aspiring to understand the universe. At the same time, to adhere to quantum mechanics was not virtuous, as it did not deliver 'understanding'. Einstein confided to a young fellow critic of the quantum theory, David Bohm, that "I believe that [the unified] laws are logically simple and that the faith in this logical simplicity is our best guide, in the sense that it suffices to start from relatively little empirical knowledge. If nature is not arranged in a way corresponding to this belief, then there is no hope at all to arrive at a deeper *understanding*."[19] As we saw, without such understanding, man will remain bound to the confines of the superficial and personal.

The close relation between religious sensibility, and moral and epistemic virtue was expressed well by Spinoza in Einstein's assessment: "those individuals to whom we owe the great creative achievements of science were all of them imbued with the truly religious conviction that this universe of ours is something perfect and susceptible to the striving for knowledge. If this conviction had not been a strongly emotional one and if those searching for knowledge had not been inspired by Spinoza's *Amor Dei Intellectualis*, they would hardly have been capable of that untiring devotion which alone enables man to attain his greatest achievements" (Einstein [1948], 56–57). Einstein insisted on the devotion of science's practitioners: He held that "the cosmic religious feeling" was experienced when the "futility of human desires and the sublimity and marvelous order which reveals itself both in nature and in the world of thought" is recognized (Einstein [1930b], p. 41). Just as Spinoza, Einstein saw personal transcendence to take place in intellectual study of nature, while thought and nature overlapped: "I have found no better expression for 'religious' than confidence in the rational nature of reality as it is accessible to human reason. Wherever this feeling is absent, science degenerates into uninspired

[18] Einstein in unpublished essay document, dated 1931. Albert Einstein Archives, Jerusalem, entry no. 2–110; emphasis as in original.

[19] Einstein to David Bohm, 24 November 1954, on pp. 181–182 in van Dongen 2010a.

empiricism."[20] So, mathematical unification stands to quantum mechanical empiricism as 'religious' stands to 'uninspired'. Einstein considered himself religious in a way that resonated with both Spinoza and his unification program: "my feeling is insofar religious as I am imbued with the consciousness of the insufficiency of the human mind to understand deeply the harmony of the Universe which we try to formulate as 'laws of nature'"; to attain such a "consciousness", "humility" was required.[21] Only when the "futility of human desires" is acknowledged will we "experience the universe as a single, significant whole"—then, moral, epistemic and personal fulfilment may be attained.

## Conclusion

Remarkably, Spinoza is not mentioned in Einstein's *Autobiographical Notes*. Still, the ideas expressed there exhibit the same interdependence of striving for unification in physics, moral and epistemic virtue, and personal deliverance, while moral and epistemic qualities particularly met in the scientist's religious gaze. At the same time, Einstein rewrote his own history and crafted his own version of his persona to justify and promote his unification attempts: theoretical and personal virtues here mirrored one another.

It need not surprise us, then, that scholars involved in quantum mechanics and its successors, when rejecting Einstein's unified field theories, not only rejected the science but also his ideas of what a theorist was supposed to be, along with his own embodiment of those ideas. Robert Oppenheimer, for instance, deemed the older Einstein as someone who had "lost contact with the profession of physics" and as "completely cuckoo." Einstein would have been "wasting his time", Oppenheimer found (Schweber 2008, 276, 279; Bird and Sherwin 2006, 64). At stake in these judgements was what it meant to be a theorist, and what a theorist's relation to the empirical is.

Clearly, it is indeed in efforts to understand the formation of disciplinary identity that personae and epistemic virtues are necessary as historiographical tools, just as they are most useful in understanding intra-disciplinary debates over future directions, as Oppenheimer's responses illustrate.

What we particularly wished to argue here is that concerns over what it meant to be a theorist were reflected in the theories constructed: a 'superficial' scholar, in Einstein's judgement, aimed for the empirical and could be content with a theory that was only concerned with probabilities and not full reality. Humble, devoted and "true" theorists, on the other hand, produced theories that were unified and mathematically natural. They were not distracted by "the immediately given."[22] The

[20] Einstein to Maurice Solovine, 1 January 1951 (Einstein 1956, 102–104).

[21] Einstein to B[?] F[?], 17 December 1952, quoted as in Jammer 1999, 122; see also Holton 2005.

[22] "Immediately given" is a term used by Einstein, see e.g. his letter to Moritz Schlick, 21 May 1917 (Schulmann et al. 1998, 456–457).

example of Einstein shows that tracking idealized 'virtuous theorists' helps to identify histories of theoretical virtues, and how they operated in theory choice. We further saw that virtue is at the same time morally and epistemically charged.

It has been expressed by philosophers of science, in frustration with the contingencies emphasized by historiography, that Thomas Kuhn's account of the role of values in theory choice should be cleansed of words with moral connotations—that instead of 'virtues' and 'values' a morally neutral and less historically contingent term as epistemic 'conditions' should be preferred when describing theory choice.[23] Yet, when we break the connection between theory choice, epistemic virtues and personae, we risk to lose sight of the people in our histories, particularly as scholars express and argue their perspective through morally charged, idealized biography. Einstein's example, at least, makes clear that Kuhn's values in theory choice should be thoroughly historicized: only then can we unearth the interweaving of epistemic and moral virtues and their role played in assessing theory.

## Bibliography

### *Primary Sources*


Born, Max, Hedwig Born and Albert Einstein. [1971] 2005. *The Born-Einstein Letters*. New York: Macmillan.

Buchwald, Diana, Tilman Sauer, Ze'ev Rosenkranz, József Illy, Virginia I. Holmes, Jeroen van Dongen, Daniel J. Kennefick, and A.J. Kox. 2006. *The Collected Papers of Albert Einstein, Volume 10: The Berlin Years, Correspondence May–December 1920 and Supplementary Correspondence, 1909–1920*. Princeton: Princeton University Press.

Einstein, Albert. 1905. Zur Elektrodynamik bewegter Körper. *Annalen der Physik* 17: 891–920.

———. [1918] 1994. Principles of Research. *In Ideas and Opinions*, 244–248. New York: The Modern Library.

———. [1930a] 1994. Johannes Kepler. In *Ideas and Opinions*, 286–291. New York: The Modern Library.

———. [1930b] 1994. Religion and Science. In *Ideas and Opinions*, 39–43. New York: The Modern Library.

———. [1931] 1994. Maxwell's Influence on the Evolution of the Idea of Physical Reality. In *Ideas and Opinions*, 291–295. New York: TheModern Library.

———. [1933] 1994. On the Method of Theoretical Physics. In *Ideas and Opinions*, 296–303. New York: The Modern Library.

———. [1934] 1994. *Good and Evil. Ideas and Opinions, 13*. New York: The Modern Library.

———. 1936. Physik und Realität. *Journal of the Franklin Institute* 221: 313–347.

———. [1948] 1994. Religion and Science: Irreconcilable? In *Ideas and Opinions*, 53–57. New York: The Modern Library.

———. [1949] 1997. Autobiographisches [and translation: Autobiographical Notes]. In *Albert Einstein: Philosopher-Scientist*, ed. Paul A. Schilpp, 1–94. La Salle: Open Court.

———. 1950. On the Generalized Theory of Gravitation. *Scientific American* 182: 13–17.


[23] This point was debated at the excellent conference 'Theoretical Virtues in Theory Choice', 12–14 July 2012, at the Zukunftskolleg of the University of Konstanz.

———. 1956. *Lettres à Maurice Solovine*. Paris: Gauthier-Villars.
Heisenberg, Werner. [1974] 1989. Encounters and Conversations with Albert Einstein. In *Encounters with Einstein and Other Essays on People, Places and Particles*, 107–122. Princeton: Princeton University Press.
Millikan, Robert A. 1949. Albert Einstein on His Seventieth Birthday. *Reviews of Modern Physics* 21: 343–345.
Schilpp, Paul A. [1949] 1997. Preface. In *Albert Einstein: Philosopher-Scientist*, ed. Paul A. Schilpp, xiii–xvi. La Salle: Open Court.
Schulmann, Robert, A.J. Kox, Michel Janssen, and József Illy. 1998. *The Collected Papers of Albert Einstein, Volume 8: The Berlin Years, Correspondence, 1914–1917*. Princeton: Princeton University Press.
Speziali, Pierre, ed. 1979. *Albert Einstein, Michele Besso. Correspondance 1903–1955*. Paris: Hermann.
Stachel, John, David C. Cassidy, Jürgen Renn, Robert Schulmann, Don Howard, and A.J. Kox. 1989. *The Collected Papers of Albert Einstein, Volume 2: The Swiss Years, Writings: 1900–1909*. Princeton: Princeton University Press.

## Secondary Sources

Bird, Kai, and Martin J. Sherwin. 2006. *American Prometheus: The Triumph and Tragedy of J. Robert Oppenheimer*. New York: Vintage.
Daston, Lorraine. 2008. A Short History of Einstein's Paradise Beyond the Personal. In *Einstein for the 21st Century: His Legacy in Science, Art, and Modern Culture*, ed. Peter Galison, Gerald Holton, and Silvan Schweber, 15–26. Princeton: Princeton University Press.
Daston, Lorraine, and Peter Galison. 2007. *Objectivity*. New York: Zone Books.
Daston, Lorraine, and Otto Sibum. 2003. Introduction: Scientific Personae and Their Histories. *Science in Context* 16: 1–8.
Gearheart, Clayton. 2002. Planck, the Quantum, and the Historians. *Physics in Perspective* 4: 170–215.
Gregory, Frederick. 2000. The Mysteries and Wonders of Natural Science: Aaron Bernstein's *Naturwissenschaftliche Volksbücher* and the Adolescent Einstein. In *Einstein: The Formative Years, 1879–1909*, ed. Don Howard and John Stachel, 23–42. Boston: Birkhäuser.
Holton, Gerald. 1968. Mach, Einstein and the Search for Reality. *Daedalus* 97: 636–673.
———. 1969. Einstein, Michelson, and the 'Crucial' Experiment. *Isis* 60: 133–197.
———. 1998. Thematic Presuppositions and the Direction of Scientific Advance. In *The Advancement of Science, and its Burdens*, 3–27. Cambridge, MA: Harvard University Press.
———. 2005. Einstein's Third Paradise. In *Victory and Vexation in Science. Einstein, Bohr, Heisenberg and Others*, 3–15. Cambridge, MA: Harvard University Press.
Howard, Don. 1993. Was Einstein Really a Realist? *Perspectives on Science* 1: 204–251.
———. 1994. Einstein, Kant, and the Origins of Logical Empiricism. In *Language, Logic, and the Structure of Scientific Theories*, ed. Wesley Salmon and Gereon Wolters, 45–105. Pittsburgh: University of Pittsburgh Press & Konstanz: Universitätsverlag.
———. 1998. Astride the Divided Line: Platonism, Empiricism, and Einstein's Epistemological Opportunism. In *Idealization in Contemporary Physics*, ed. Niall Shanks, 143–163. Amsterdam and Atlanta: Rodopi.
Jammer, Max. 1999. *Einstein and Religion*. Princeton: Princeton University Press.
Janssen, Michel, and Jürgen Renn. 2007. Untying the Knot: How Einstein Found His Way Back to Field Equations Discarded in the Zurich Notebook. In *The Genesis of General Relativity, Volume 2: Einstein's Zurich Notebook: Commentary and Essays*, ed. M. Janssen, J. Norton, J. Renn, T. Sauer, and J. Stachel, 839–925. Dordrecht: Springer.

Jungnickel, Christa, and Russell McCormmach. 1986. *Intellectual Mastery of Nature: Theoretical Physics from Ohm to Einstein*. Vol. 2 vols. Chicago: University of Chicago Press.

Kuhn, Thomas S. 1977. Objectivity, Value Judgment and Theory Choice. In *The Essential Tension*, 320–339. Chicago: University of Chicago Press.

———. 1978. *Black Body Theory and the Quantum Discontinuity, 1894–1912*. Oxford: Clarendon Press.

Nadler, Steven. 2001. *Baruch Spinoza*. Stanford Encyclopedia of Philosophy. https://plato.stanford.edu/entries/spinoza/. Accessed 23 December 2016.

Norton, John. 1984. How Einstein Found His Field Equations, 1912–1915. *Historical Studies in the Physical Sciences* 31: 231–316.

Renn, Jürgen, and Tilman Sauer. 2007. Pathways Out of Classical Physics: Einstein's Double Strategy in His Search for the Gravitational Field Equation. In *The Genesis of General Relativity, Volume 1: Einstein's Zurich Notebook: Introduction and Source*, ed. M. Janssen, J. Norton, J. Renn, T. Sauer, and J. Stachel, 113–312. Dordrecht: Springer.

Sauer, Tilman. 2006. Field Equations in Teleparallel Space-time: Einstein's Fernparallelismus Approach Toward Unified Field Theory. *Historia Mathematica* 33: 399–439.

———. 2014. Einstein's Unified Field Theory Program. In *The Cambridge Companion to Einstein*, ed. Michel Janssen and Christoph Lehner, 281–305. Cambridge: Cambridge University Press.

Schweber, Silvan S. 2008. *Einstein and Oppenheimer: The Meaning of Genius*. Cambridge, MA: Harvard University Press.

Tai, Chaokang. 2017. Left Radicalism and the Milky Way: Connecting the Scientific and Socialist Virtues of Anton Pannekoek. *Historical Studies in the Natural Sciences* 47: 200–254.

Tai, Chaokang, and Jeroen van Dongen. 2016. Anton Pannekoek's Epistemic Virtues in Astronomy and Socialism: Personae and the Practice of Science. *Low Countries Historical Review* 131 (4): 55–70.

Van Dongen, Jeroen. 2002. Einstein and the Kaluza-Klein Particle. *Studies in History and Philosophy of Modern Physics* 33: 185–210.

———. 2004. Einstein's Methodology, Semivectors and the Unification of Electrons and Protons. *Archive for History of Exact Sciences* 58: 219–254.

———. 2007. Emil Rupp, Albert Einstein, and the Canal Ray Experiments on Wave-Particle Duality: Scientific Fraud and Theoretical Bias. *Historical Studies in the Physical and Biological Sciences* 37: 73–120.

———. 2009. On the Role of the Michelson-Morley Experiment: Einstein in Chicago. *Archive for History of Exact Sciences* 63: 655–663.

———. 2010a. *Einstein's Unification*. Cambridge: Cambridge University Press.

———. 2010b. On Einstein's Opponents, and Other Crackpots. *Studies in History and Philosophy of Modern Physics* 41: 78–80.

Wazeck, Milena. 2009. Einsteins Gegner. In *Die öffentliche Kontroverse um die Relativitätstheorie in den 1920er Jahren*. Frankfurt a.M.: Campus.

Wertheim, David J. 2011. *Salvation through Spinoza: A Study of Jewish Culture in Weimar Germany*. Leiden: Brill.

# Scholarly Vices: Boundary Work in Nineteenth-Century Orientalism

**Christiaan Engberts and Herman Paul**

**Abstract** What are epistemic vices? Drawing on the cases of two mid-nineteenth-century Orientalists, both of whom were accused of serious vices, this chapter argues that dispositions perceived as detrimental to scholarly work were often difficult to distinguish from social or religious vices. Indeed, when Heinrich Ewald (1803–1875) and Reinhart Dozy (1820–1883) were blamed for "dogmatism" and "rashness," these were epistemic as much as social and religious vices. This chapter therefore proposes to exchange the concept of epistemic vices for the more inclusive category of scholarly vices. What this alternative phrase seeks to convey is that, for at least some nineteenth-century academics, the pursuit of epistemic aims was inseparable from meeting social expectations, engaging in political projects, and fulfilling religious duties. The relative importance of these various aims, however, was contested. Accordingly, criticism of Dozy and Ewald was part of a broader debate on the scholarly vocation and not seldom amounted to a form of "boundary work" between competing conceptions of Orientalist scholarship.

**Keywords** Epistemic virtues • Epistemic vices • Scholarly vices • Boundary work • Scholarly personae • History of humanities • Orientalism • History of Orientalism • Heinrich Ewald • Reinhart Dozy

## Introduction

While language of virtue ("impartiality," "honesty," "accuracy") such as used by nineteenth and twentieth-century scholars is increasingly receiving attention from historians of science interested in historicizing the concepts and categories that academics and non-academics alike invoke in articulating their understanding of the scholarly vocation, language of vice ("bias," "prejudice," "sloppiness") has so far been studied surprisingly little. Although Sari Kivistö in her stimulating book on early modern "vices of learning" has shown conclusively that language of vice

C. Engberts (✉) • H. Paul
Institute for History, Leiden University, Leiden, The Netherlands
e-mail: c.a.engberts@hum.leidenuniv.nl; h.j.paul@hum.leidenuniv.nl



J. van Dongen, H. Paul (eds.), *Epistemic Virtues in the Sciences and the Humanities*, Boston Studies in the Philosophy and History of Science 321,
https://doi.org/10.1007/978-3-319-48893-6_6

played crucial roles in defining and defending ideals of scholarly behavior (Kivistö 2014), empirical case studies on the meanings and functions of vices in scholarly debates, reviews, or textbooks hardly exist. Ian Hesketh's analysis of "errors" attributed to the Victorian historian James Anthony Froude is one of only very few studies demonstrating that norms or expectations regarding scholarly conduct can be inferred at least as well from what scholars reject in terms of vices as from what they praise in terms of virtues (cf. Trüper forthcoming).

If this chapter follows Hesketh's lead by examining what it meant for the Orientalists Heinrich Ewald (1803–1875) and Reinhart Dozy (1820–1883) to be accused of vices such as "dogmatism" and "rashness," it does so in the light of a specific conceptual question. Historians of nineteenth and twentieth-century virtue language often use the modern category of "epistemic virtues" to highlight that the virtues under discussion were not simply moral ones, but (also) were supposed to serve such epistemic goals as knowledge and understanding of reality. "Objectivity," for instance, can be called an epistemic virtue to the extent that it was believed to increase the epistemic reliability of scholarly research (Daston and Galison 2007; Daston 2014). Our question is to what extent the contrast category of "epistemic vices," such as employed by Quassim Cassam (2016) and Ian James Kidd (2016), can serve as a prism for studying language of vice. Did the vices with which Ewald and Dozy found themselves charged possess sufficient epistemic relevance to warrant classification as epistemic vices?

As long as epistemic vices are equated with "cognitive defects," or character traits impeding with "effective and responsible inquiry" (Cassam 2016: 160), the answer must be negative. For although some of the vices attributed to Ewald and Dozy were epistemic in the sense of hampering serious scholarly research, most of the criticism they elicited focused on social and religious faults. Ewald's "dogmatism," for example, was perceived as hindering the pursuit of epistemic aims, but also, at the same time, as complicating teacher-student relations and as frustrating collegial cooperation. A single vice, then, could have multiple layers of meaning, of which the epistemic was not necessarily the most important one. This chapter therefore closes with suggesting that the adjective "epistemic" might be too restrictive to capture the meanings associated with Ewald's and Dozy's perceived vices. The more inclusive category of "scholarly vices" might better convey that, for nineteenth-century scholars at least, the pursuit of epistemic aims was often inseparable from meeting social expectations, engaging in political projects, and fulfilling religious duties.[1]

## Reinhart Dozy

Reinhart Dozy, a Leiden-based Arabic scholar widely known for his *Histoire des Musulmans d'Espagne* (1861), among other publications (Brugman 1989; Vrolijk and Van Leeuwen 2014: 95–102), offers a case in point. He became the center of a

[1] Much of what follows draws and expands on Paul 2015 and Engberts 2016.

controversy after the appearance of his study, *De Israëlieten te Mekka* (The Israelites in Mecca) (1864). Scholars near and abroad had awaited the book with anticipation, if only because Dozy, not otherwise known for boldness or impudence, had announced that it would place the origins of Islam in an entirely new light: "I have obtained a result that ... is entirely at odds with reigning views; it sounds very strange, and given that in the realm of scholarship no one can demand to be taken on his word, it requires a lengthy, scholarly demonstration of proof" (Dozy 1863: 2). In private correspondence, colleagues throughout Europe had speculated how the argument might look like (e.g., Theodor Nöldeke to Michaël Jan de Goeje, 1 May 1864; Abraham Geiger to Reinhart Dozy, 16 June 1864). But none had foreseen how spectacular the result would be. *De Israëlieten te Mekka* argued that the Kaaba stone in Mecca had originally been a stone god worshipped by the Israelites before their conversion to monotheism. Abraham had been the name of this stone. Only in the days of Ezra, pious Israelite scribes had transformed Abraham into the patriarch known from the book of Genesis. The Meccan festival in turn had been established by the tribe of Simeon after the model of Baal feasts in ancient Canaan. So, for Dozy, Islam stemmed from Israelite sources, with the Israelites in those days being primitive stone worshippers (Dozy 1864).

Although Michaël Jan de Goeje, Dozy's closest pupil, in near-religious language welcomed the book as "a series of revelations regarding the darkest and most enigmatic issues" in the history of Islam (De Goeje 1864: 298), *De Israëlieten te Mekka* caused a rather predictable storm of controversy. Lengthy reviews, often critical in tone, appeared in scholarly journals as well as in the popular press, while pamphlets attacking or defending Dozy were published by friends and foes alike. As even Karl Marx reported in an 1864 letter to Friedrich Engels, "the book has caused a great uproar ... particularly since Dozy is the most learned Dutch Orientalist – and a professor in Leiden to boot!" (Marx 1979: 186, translation slightly modified).

Given the prominent position of categories of virtue and vice in nineteenth-century moral discourse (Hettling and Hoffmann 2000), it comes as no surprise that both sympathizers and critics evaluated Dozy's work in terms of virtues and vices. The two virtues most frequently ascribed to Dozy were "shrewdness" (*scherpzinnigheid*) and "frankness" (*vrijmoedigheid*). Friendly and skeptically inclined reviewers alike agreed that Dozy had a "sharp eye," which enabled him to excel in "criticism" (Matthes 1864: 351; De Khanikoff 1864: 433). Whether Dozy's hypotheses were correct or not, no one questioned that they could provide a "salutary shock" and stimulate fresh research (Kuenen 1870: 406). Critics, by contrast, accused Dozy of "rashness" (*lichtvaardigheid*) or even "levity" (*lichtzinnigheid*) in advancing far-reaching hypotheses on the mere basis of speculative inferences. Also, they deemed him guilty of "prejudice" and "dogmatism" because of his tendency to interpret difficult passages in such a way as to make them fit his argument. Dozy's "love for his theory," in other words, surpassed his love of truth (Van Herwerden 1865: 395; Land 1864: 229). Finally, what some perceived as "frankness" appeared as "insolence" (*aanmatiging*) or "impertinence" (*brutaliteit*) in the eyes of others. How could one treat such sensitive topics as ancient Jewish and Muslim religion in such an iconoclastic manner, without even considering the possibility of alternative readings and other, more nuanced, approaches (Elte 1871: 30)? The German Old Testament scholar Karl Heinrich Graf spoke for many when he concluded that Dozy's research, though carried out with "acuteness and learning," had a scholarly worth of "nothing, absolutely nothing":

> Everyone else would regard the results that are presented here with great confidence as indisputably proven historical facts as a web of light-hearted hypotheses, daring and adventurous inferences and conjectures, and be surprised at the rashness with which the most far-reaching conclusions are often drawn from unproven assumptions… Everything floats in the air, nothing rests on a firm, scholarly smoothed ground, everywhere the lack of solid critical rules is manifested in a most strident manner (Graf 1865: 350).

If the deficiencies of Dozy's book were so obvious, why did critics spend ample amounts of time and ink exposing them in public? As Marx in his above-quoted letter perceptively observed, part of the scandal was that Dozy was a well-known professor with a solid reputation in his field. Ironically, he was known for precisely those virtues that *De Israëlieten te Mekka* seemed to lack: conscientiousness, cautiousness, and philological precision. His research monographs and Arabic source editions testified to what has been called a "philological ethos," characterized by carefulness, attentiveness to details, and painstaking source criticism (Kolk 1989; Paul 2012). Given Dozy's good reputation, some critics assumed that *De Israëlieten te Mekka* could not be entirely wrong-headed (Weil 1864: 596). Others criticized Dozy in public precisely because of this reputation – his good name could give inappropriate credence to the message of his Mecca book – or expressed surprise that "a man of such first-rate manly learning" could produce such "wild, youthful leaps and bounds" (Hofstede de Groot 1865: 373).

More important, though, is that quite a few critics perceived Dozy and his book as typical products of Leiden's liberal theological faculty, which was internationally known as a bulwark of historical criticism. It was here that Abraham Kuenen developed his documentary hypothesis about the sources of the Pentateuch (the first five books of the Hebrew Bible traditionally attributed to Moses). Also, it was here that entire cohorts of "modern" or "liberal" Protestant pastors received their training (Dirksen and Van der Kooij 1993; Buitenwerf-van der Molen 2007). Accordingly, it was not only Dozy's personal reputation that was at stake in defending or criticizing *De Israëlieten te Mekka*. When colleagues like Kuenen and former students such as J.P.N. Land, J.C. Matthes, and Hendrik Oort rushed to Dozy's defense, they did so, among other things, to defend the good right of historical criticism of the Bible and, more specifically, the legitimacy for Old Testament scholars to challenge traditional religious viewpoints (Land 1864; Matthes 1864; Oort 1864). Conservative Jewish critics, to the contrary, did not hesitate to extrapolate their judgments on Dozy to the "Leiden school" in general:

> Professor *Dozy* writes "The Israelites in Mecca" and all liberal theologians, headed by Prof. *De Goeje*, look on amazed and struck dumb by those heavenly revelations from *Sinai*, no at *Leiden*. They are inexhaustible in their cheers, they dance and shout with joy like the Egyptians when Apis honors them with a visit. *Kuenen* writes "The Religion of the Israelites" and the entire choir of liberal theologians sings Hallelujah, yea is seized with convulsion caused by joy. So it is hardly possible for our pastors not to vow on the oracles of the great masters who, for the benefit of their loyal, credulous disciples, are busily brewing brand new revelations (N.N. 1871).

Not only Jewish critics seized the opportunity to show what was wrong with historical criticism in general and Leiden liberalism in particular. The Netherlands Reformed pastor Lambertus Tinholt, too, used *De Israëlieten te Mekka* as an occasion for ridiculing historical criticism with a parody that relegated Charles V to the realm of myth, complete with phrases that could have walked straight out of Dozy's book ("These assertions are strange, but I will demonstrate that they are not unfounded" [Tinholt 1864: 416]). Likewise, a colleague of Tinholt translated Graf's highly critical review into Dutch in order to contribute to what he called a "battle" against "a falsely named scholarship," intend on assaulting Scripture (Leenmans 1864: iv). Ran HaCohen's observation that Dozy's orthodox Jewish critics were driven by religious concerns at least as much as by epistemic ones (HaCohen 2010: 145) can therefore be extended to a great number of Dozy's commentators, defenders and critics of Leiden-style liberalism included. For most of the parties involved, epistemic and religious agendas were rather intertwined.

Therefore, in answer to the question what it meant for Dozy to be accused of vicious scholarly conduct, three observations can be made. One is that nineteenth-century Orientalists did not share an undisputed catalog of virtues and vices. Whether it was a virtue or a vice to challenge traditional religious views, as Dozy did by portraying the ancient Israelites as stone worshippers, largely depended on one's attitudes towards the Bible or on one's position on the spectrum between theological orthodoxy and liberalism. Secondly, something like guilt by association existed. Dozy's vices were perceived as dangerous, and therefore as worth combatting, to the extent that they were seen as corresponding to the vices of Protestant liberalism. There is a sense, therefore, in which Dozy's shortcomings were supra-individual: they were labeled as vices because they were seen as serving liberal theological purposes. This, finally, implies that Dozy's "rashness" and "impertinence" were religious vices as much as epistemic ones. If religious considerations helped determine what counted as proper epistemic conduct, religious and epistemic vices cannot be distinguished unambiguously.

## Heinrich Ewald

If epistemic vices did not exist in isolation from religious ones, they could neither be separated from moral or, as we would like to call them, social ones. This is evidenced by the example of Heinrich Ewald, a Göttingen Orientalist who was among the first to accuse Dozy of a "vast number of continuously new and continuously more far-reaching errors" (Ewald 1864: 1270). Like Dozy, Ewald was a man of many talents, whose erudition and linguistic skills in particular were widely respected. He was an influential teacher, too, who counted such future heavyweights as August Dillmann, Theodor Nöldeke, and Julius Wellhausen among his students (Davies 1903; Perlitt 1987). Yet, at the same time, he was a near-permanent object of scorn and gossip among his colleagues.

One factor contributing to his bad reputation was Ewald's refusal to adopt a model of scholarship that separated academic learning from religious and political edification. While most German Orientalists at the time followed Heinrich Leberecht Fleischer in identifying Orientalist scholarship with philological *Kleinarbeit* (Mangold 2004: 78–115), Ewald never lost sight of bigger questions. As one of the Göttingen Seven, who had lost their chairs in 1837 after protesting against the King of Hanover (Lampe 2002: 91–97), Ewald was a staunch anti-Prussian who preferred not to hide his political views. In his extended polemics, political, theological, and philological arguments permeated his philological work (Perlitt 1987: 168–169). Notably, his main argument against Biblical scholarship intend on separating "historical facts" from "myths" was that it would favor an "ultramontane" approach to Scripture. As he explained to a German philologist: "The Roman Church profits from seeing nothing but mysteries in the Bible. If we Protestants concede that a Biblical book is nothing but a book of mysteries, we thereby surrender ourselves" (Ewald to Friedrich Zarncke, 8 September 1851).

More problematic than this anti-Catholic stance, which was widely shared among the German Protestant professorate, was the harsh and *ad hominem* tone in which Ewald used to conduct his polemics. Typical was his outburst against the Leipzig theologian Franz Delitzsch: "Immediately, at his first appearance in Leipzig, this renowned Erlanger theologian behaved so hostile and conceited against all that belongs to conscientious Biblical scholarship, and, when you look at what is essential, he shows through all the contractions in which he loses himself still so little of any serious scholarship, that he will easily be found out by all who possess only a little bit more true love for truth and certainty" (Ewald 1865: 35–36). Speaking about Leipzig, Ewald seized the opportunity also to lash out at Fleischer, "who, I hope, does not want to betray Christianity as well as the Evangelical faith" (Ewald 1865: 37). These reproaches continued earlier polemics against Fleischer, also in private correspondence (as recounted by Fleischer in a letter to Nöldeke, 9 August 1863).

His thundering voice and *ad hominem* charges earned Ewald a dubious reputation among colleagues. It was Ewald's prose style, among other things, that led Wellhausen to characterize his teacher as "congenial" with the Old Testament prophets (Wellhausen 1901: 73). Fleischer in turn wrote ironically: "In the end we fully owe it to his magnanimity and forbearance only, that we quasi-scholars still exist; if he would like to destroy us, it would cost him, like the JHWH of the Old Testament, just a whiff of his mouth and we would be gone!" (Fleischer to de Goeje, 11 March 1865; see also Fleischer to Nöldeke, 2 June 1865). When colleagues, especially in private correspondence, referred to "the Old Testament prophet," "the pope of Göttingen," or "the Buddha of Göttingen," there could be little doubt whom they had in mind (De Goeje 1865: 531; Fleischer to de Goeje, 5 December 1865; Nöldeke to Ignaz Goldziher, 22 February 1904).

The vice most frequently attributed to the prophetic Göttingen professor was "dogmatism." Although dogmatism has not always been considered a vice (Kuhn 1963), nineteenth-century Orientalists generally agreed on its vicious nature: they saw dogmatism as a character trait that hampered unbiased and evenhanded research. In Ewald's case, critics complained that the Göttingen professor was too convinced of his own beliefs to pay serious attention to alternative hypotheses, which he tended either to ignore or to condemn for being inconsistent with his own superior views (N.N. 1855, N.N. 1857, N.N. 1859). As an English scholar complained: "[I]t is characteristic of Ewaldian criticism not only to deal unfairly with rival theories, but to treat whatever it has once sanctioned and adopted as irrefragable and irrevocable" (Mackay 1863: 349). Paired to the habit of considering "the celebrity of any opinion not emanating from himself" as "sufficient reason for condemning and contradicting it" (Mackay 1863: 345; approvingly quoted in Dozy to A.C. Kruseman, 6 September 1864), Ewald's dogmatism was perceived as a serious hindrance to respectable scholarship.

Yet this epistemic vice was a social one, too, in the sense that it complicated collegial relations and frustrated productive exchange of ideas, as Theodor Nöldeke, among others, experienced when working alongside Ewald as a *Privatdozent* in Göttingen. In 1861, he complained to De Goeje: "If I did not have Ewald here, whom I cannot make my enemy, I would straight out lecture on the Old Testament" and "conduct myself more freely" in matters of historical criticism (Nöldeke to de Goeje, 28 February 1861 and 27 April 1861). A visit to the manuscript room of the British Museum turned into a fiasco once Ewald discovered Ernst Meier sitting at a desk. He "couldn't sit in the same room" with a former student whom he perceived as having apostated by joining the Tübingen School around Ferdinand Christian Baur (de Goeje to Nöldeke, October 10, 1862; cf. Junginger 1999: 21). Similar anecdotes circulated among Orientalists all over Europe (e.g., Fleischer to Nöldeke, 28 March 1865; Fleischer to Friedrich Zarncke, undated), as professional gossip serving the purpose of applying and reinforcing scholarly standards of conduct.

The extent to which Ewald was perceived as crossing a line is evidenced not only by Orientalists who openly dissociated themselves from Ewald (Zarncke 1865) or cut off all contact with him (Littmann 1931: 56), but also by the fact that his name developed into shorthand for a particular type of Orientalist scholar. Fleischer, for instance, characterized Ewald's former student Ferdinand Hitzig as a "second Ewald." "This self-delusion! The loosest ideas turn into sacrosanct truths of God's mercy, because they pop up into Mr. Hitzig's skull like falling stars" (Fleischer to Nöldeke, 2 March 1872). Ewald's successor in Göttingen, Paul de Lagarde, also became known as "Ewald II" after the publication of a volume full of polemical letters and devastating book reviews:

> The Göttingen Ewald II seems to have broken off completely with the sana ratio [sound reason] after the conclusions of "Aktenstücke und Glossen aus dem deutschen Gelehrtenleben," p. 116. The megalomania expressed therein is so pronounced that the only question really is whether we should allow this madman to continue such a public scandal at the lectern or whether we should lead him away to where he belongs (Fleischer to Nöldeke, 31 December 1888).

Ewald's name, then, came to serve not only as a proper name, referring to the man himself, but also as a generic one, denoting what one might call an "unscholarly persona" or a model of scholarly conduct that most Orientalists preferred not to imitate.

Unlike Dozy, then, Ewald was not primarily attacked on religious grounds. In his case, criticism rather focused on his inability to cooperate with colleagues holding different views on issues Ewald deemed non-negotiable. What the two men had in common, though, is that they were criticized in terms of epistemic vices ("rashness," "impertinence," "dogmatism") and that this criticism was targeted not only at them, but also at kindred spirits or similarly dispositioned colleagues. Moreover, the vices in question, though epistemic at first sight, had more than epistemic connotations: their social and religious layers of meaning were at least as important. This is not to say that the epistemic idiom was a mere instrument for framing religious or social deviance in academically respectable terms. In both cases, real epistemic issues were at stake. These were intertwined, however, with additional concerns that lent additional meaning to the vices in question. So, when Robert William Mackay complained that Ewald "wraps his virtue in an obscurity of inflated verbiage" (Mackay 1863: 343) or when William Wright, speaking about Ewald, sighed that "*learning* there is plenty, God knows, but very little *discretion*" (Maier 2011: 163), this criticism invoked standards of scholarly conduct that had epistemic as well as social aspects to them.

## Conclusion

Accordingly, a first conclusion that can be drawn from these case studies is that the adjective "epistemic" must be used with care. "Epistemic" should not be taken in a restricting sense, as denoting a class of virtues neatly separated from moral, social, political, or religious virtues. As Camille Creyghton and others have argued, a weak reading of the adjective, according to which "epistemic" merely indicates which layer of meaning appears to be most relevant in the context at hand, seems more appropriate for historians of science and the humanities (Creyghton et al. 2016). In order to highlight that the vices that nineteenth-century scholars were supposed to avoid can not be classified exclusively as epistemic either, it might even be worthwhile to exchange the category of "epistemic vices" for the more inclusive concept of "scholarly vices" – with an adjective emphasizing that scholars do not pursue epistemic aims in isolation from social, moral, religious, or political ones, and consequently cannot distinguish unambiguously between their corresponding vices.

The cases of Dozy and Ewald furthermore illustrate that European Orientalists in the second half of the nineteenth century did not agree on the importance of social or religious aims in relation to epistemic ones. Many of the battles they fought over what constituted appropriate or inappropriate scholarly conduct were battles over hierarchies of aims or, as they have elsewhere been called, "constellations of commitments" (Paul 2014: 363–365; see also Kuhn 1977). How much weight did the one have in relation to the other? How important was, for instance, humility vis-à-vis Scripture compared to critical examination of ancient texts? How relevant were collegial manners in relation to truth-finding? And how important were political agendas compared to philological ones? In the polarized world of nineteenth-century Orientalism, agreement on these fundamental issues did not remotely exist (Mangold 2004; Marchand 2009; Trüper 2013).

This, finally, helps explain why Ewald and Dozy, each in their own way, evoked much critical commentary. Partly, the two were perceived as displaying qualities that critics from backgrounds as diverse as orthodox Judaism and liberal Protestantism regarded as detrimental to serious scholarship. Even if the critics anything but agreed on what good scholarship required in terms of dispositions or character traits, they were able to unite, *ex negativo*, in condemning Dozy's "speculation" and Ewald's "dogmatism." Partly also, the two were at the heart of often vigorous polemics between competing schools or approaches. At stake was not just their individual performance, but the virtues and vices of historical criticism as advocated in Leiden or Arabic philology as practiced in Leipzig. The vices ascribed to Dozy and Ewald can therefore well be interpreted as instruments for "boundary work" – attempts at defending a particular understanding of the scholarly vocation by contrasting it polemically to competing ideas of what scholarship is and how scholars should behave (Gieryn 1983; Hesketh 2008). If Dozy and Ewald became known as models of scholarly vice, these reputations were due at least in part to deliberate boundary work on the part of scholars committed to different hierarchies of aims than those embodied by, or ascribed to, the protagonists of this chapter.

**Acknowledgment** Funding was generously provided by the Netherlands Organization for Scientific Research (NWO).

## Bibliography

### *Primary Sources (Unpublished)*

Library of the Hungarian Academy of Sciences:
Theodor Nöldeke to Ignaz Goldziher, GIL/32/01.
Leiden University Library:
Reinhart Dozy to A.C. Kruseman., LTK 1505.
Heinrich Leberecht Fleischer to Michaël Jan de Goeje, BPL 2389.
Michaël Jan de Goeje to Theodor Nöldeke, BPL 2389.
Theodor Nöldeke to Michaël Jan de Goeje, BPL 2389.
Leipzig University Library:
Heinrich Ewald to Friedrich Zarncke, NL 249/1/E/1051.
Heinrich Leberecht Fleischer to Friedrich Zarncke, NL 249/1/F/409.
Tübingen University Library:
Heinrich Leberecht Fleischer to Theodor Nöldeke, Md 782 A 68.

### *Primary Sources (Published)*

de Goeje, M.J. 1864. Een stap vooruit. *De Gids* 28 (2): 297–312.
———. 1865. Bibliographisch album. *De Gids* 29 (1): 531–548.

Dozy, R. 1863. *Het islamisme*. Haarlem: A.C. Kruseman.

———. 1864. *De Israëlieten te Mekka van Davids tijd tot in de vijfde eeuw onzer tijdrekening*. Haarlem: A.C. Kruseman.

Elte, Philip. 1871. *Moderne theologie in Nederland*. Amsterdam: J.B. de Mesquita.

E[wald], H[einrich]. 1864. Review of *De Israëlieten te Mekka* by Reinhart Dozy. *Göttingische Gelehrte Anzeigen*: 1265–1280.

———. 1865. Ueber die heutige sicherheit des verständnisses und der anwendung der Bibel. *Jahrbücher der biblischen Wissenschaft* 12: 26–64.

Graf, K.H. 1865. Review of *De Israëlieten te Mekka* by Reinhart Dozy. *Zeitschrift der Deutschen Morgenländischen Gesellschaft* 19: 330–351.

[Hofstede de] G[root], [Petrus]. 1865. De oorsprong van de Mohamedaansche godsdienst. *Waarheid in Liefde*: 373–376.

Khanikoff, N. 1864. de. 1864. Review of *De Israëlieten te Mekka* by Reinhart Dozy. *Journal Asiatique* 4: 433–449.

Land, J.P.N. 1864. Een nieuwe lichtstraal op het Oude Verbond. *De Nederlandsche Spectator*: 227–229.

Leenmans, H. A. 1864. Aan den lezer. In *De Israëlieten te Mekka van dr. R. Dozy, beoordeeld door dr. K.H. Graf*. Trans. H. A. Leenmans, iii–iv. Utrecht: Kemink en Zoon.

Littmann, Enno. 1931. Theodor Nöldeke. *Nachrichten von der Gesellschaft der Wissenschaften zu Göttingen: Geschäftliche Mitteilungen aus dem Berichtsjahr* 1930 (31): 48–57.

Mackay, R.W. 1863. *The Tübingen School and Its Antecedents: A Review of the History and Present Condition of Modern Theology*. Edinburgh: Williams and Norgate.

Marx, Karl. 1979. *The Letters of Karl Marx*. Trans. Saul K. Padover. Englewood Cliffs: Prentice-Hall.

Matthes, J.C. 1864. Eene nieuwe ontdekking. *De Tijdspiegel II*: 349–369.

N. N. 1855. Geschichte Christus' und seiner Zeit. *Literarisches Centralblatt* (31 March): 197–198.

———. 1857. Die Sendschreiben des Apostels Paulus übersetzt und erklärt. *Literarisches Centralblatt* (4 July): 417–419.

———. 1859. Geschichte des Apostolischen Zeitalters bis zur Zerstörung Jerusalems. *Literarisches Centralblatt* (15 January): 33–36.

———. 1871. Beoordeling der recensie van het Handboek voor de geschiedenis der Joden door Sluijs en Hoofiën in het Leeskabinet. In *Nieuw Israëlietisch Weekblad* (28 July).

Oort, H. 1864. *De dienst der Baälim in Israël: naar aanleiding van het geschrift van dr. R. Dozy 'De Israëlieten te Mekka.'* Leiden: P. Engels.

Tinholt, L. 1864. Eene Jakoetische voorlezing uit de 38ste eeuw onzer jaartelling. *Stemmen voor Waarheid en Vrede* 1: 411–444.

van Herwerden, C.H. 1865. Over de steenen- en boomendienst van Israël. *Waarheid in Liefde* 1: 377–398.

[Weil], [Gustav]. 1864. Review of *De Israëlieten te Mekka* by Reinhart Dozy. *Heidelberger Jahrbücher der Literatur* 57: 595–602.

Wellhausen, Julius. 1901. Heinrich Ewald. In *Festschrift zur Feier des hundertfünfzigjährigen Bestehens der Königlichen Gesellschaft der Wissenschaften zu Göttingen: Beiträge zur Gelehrtengeschichte Göttingens*, 61–88. Berlin: Weidmannsche Buchhandlung.

[Zarncke], [Friedrich]. 1865. Anfrage. *Literarisches Centralblatt* (9 December): 1375–1376.

## Secondary Literature

Brugman, J. 1989. Dozy, a Scholarly Life According to Plan. In *Leiden Oriental Connections, 1850–1940*, ed. Willem Otterspeer, 62–81. Leiden: Brill.

Buitenwerf-van der Molen, Mirjam. 2007. *God van vooruitgang: de popularisering van het modern-theologische gedachtegoed in Nederland, 1857–1880*. Verloren: Hilversum.

Cassam, Quassim. 2016. Vice Epistemology. *The Monist* 99: 159–180.
Creyghton, Camille, et al. 2016. Virtue Language in Historical Scholarship: The Cases of Georg Waitz, Gabriel Monod and Henri Pirenne. *History of European Ideas* 42: 924–936.
Daston, Lorraine. 2014. Objectivity and Impartiality: Epistemic Virtues in the Humanities. In *The Making of the Humanities*, ed. Rens Bod, Jaap Maat, and Thijs Weststeijn, vol. 3, 27–41. Amsterdam: Amsterdam University Press.
Daston, Lorraine, and Peter Galison. 2007. *Objectivity*. New York: Zone Books.
Davies, Thomas Witton. 1903. *Heinrich Ewald: Orientalist and Theologian: 1803–1903: A Centenary Appreciation*. London: T. Fisher Unwin.
Dirksen, P.B., and A. van der Kooij, eds. 1993. *Abraham Kuenen (1828–1891): His Major Contributions to the Study of the Old Testament: A Collection of Old Testament Studies Published on the Occasion of the Centenary of Abraham Kuenen's Death (10 December 1991)*. Leiden: Brill.
Engberts, Christiaan. 2016. Gossiping about the Buddha of Göttingen: Heinrich Ewald as an Unscholarly Persona. *History of Humanities* 1: 371–385.
Gieryn, Thomas F. 1983. Boundary-Work and the Demarcation of Science from Non-Science: Strains and Interests in Professional Ideologies of Scientists. *American Sociological Review* 48: 781–795.
HaCohen, Ran. 2010. *Reclaiming the Hebrew Bible: German-Jewish Reception of Biblical Criticism*. Trans. M. Engel. Berlin: Walter de Gruyter.
Hesketh, Ian. 2008. Diagnosing Froude's Disease: Boundary Work and the Discipline of History in Victorian Britain. *History and Theory* 47: 373–395.
Hettling, Manfred, and Stefan-Ludwig Hoffmann, eds. 2000. *Der bürgerliche Wertehimmel: Innenansichten des 19. Jahrhunderts*. Göttingen: Vandenhoeck & Ruprecht.
Junginger, Horst. 1999. *Von der philologischen zur völkischen Religionswissenschaft: Das Fach Religionswissenschaft an der Universität Tübingen von der Mitte des 19. Jahrhunderts bis zum Ende des Dritten Reiches*. Franz Steiner: Stuttgart.
Kidd, Ian James. 2016. Charging Others with Epistemic Vice. *The Monist* 99: 181–197.
Kivistö, Sari. 2014. *The Vices of Learning: Morality and Knowledge at Early Modern Universities*. Leiden: Brill.
Kolk, Rainer. 1989. Wahrheit, Methode, Charakter: Zur wissenschaftlichen Ethik der Germanistik im 19. Jahrhundert. *Internationales Archiv für Sozialgeschichte der deutschen Literatur* 14: 50–73.
Kuenen, A. 1870. Critische bijdragen tot de geschiedenis van den Israëlietischen godsdienst. *Theologisch Tijdschrift* 4: 391–426.
Kuhn, Thomas S. 1963. The Function of Dogma in Scientific Research. In *Scientific Change: Historical Studies in the Intellectual, Social and Technological Conditions for Scientific Discovery and Technical Innovation, from Antiquity to the Present*, ed. A.C. Crombie, 347–369. London: Heinemann.
———. 1977. Objectivity, Value Judgment, and Theory Choice. In *The Essential Tension: Selected Studies in Scientific Tradition and Change*, 320–339. Chicago: University of Chicago Press.
Lampe, Jörg H. 2002. Die 'Göttinger Sieben' und der Verfassungskampf von 1837 bis 1841. In *Göttingen: Geschichte einer Universitätsstadt*, ed. Ernst Böhme and Rudolf Vierhaus, 91–97. Göttingen: Vandenhoeck & Ruprecht.
Maier, Bernhard. 2011. *Semitic Studies in Victorian Britain: A Portrait of William Wright and His World through His Letters*. Würzburg: Ergon.
Mangold, Sabine. 2004. *Eine 'weltbürgerliche Wissenschaft': Die deutsche Orientalistik im 19. Jahrhundert*. Stuttgart: Franz Steiner.
Marchand, Suzanne L. 2009. *German Orientalism in the Age of Empire: Religion, Race, and Scholarship*. Cambridge: Cambridge University Press.
Paul, Herman. 2012. The Scholarly Self: Ideals of Intellectual Virtue in Nineteenth-Century Leiden. In *The Making of the Humanities*, ed. Rens Bod, Jaap Maat, and Thijs Weststeijn, vol. 2, 397–411. Amsterdam: Amsterdam University Press.
———. 2014. What Is a Scholarly Persona? Ten Theses on Virtues, Skills, and Desires. *History and Theory* 53: 348–371.

———. 2015. Virtue Language in Nineteenth-Century Orientalism: A Case Study in Historical Epistemology. *Modern Intellectual History*. doi.org/10.1017/S1479244315000293.

Perlitt, Lothar. 1987. Heinrich Ewald: Der Gelehrte in der Politik. In *Theologie in Göttingen: Eine Vorlesungsreihe*, ed. Bernd Moeller, 157–212. Göttingen: Vandenhoeck & Ruprecht.

Trüper, Henning. 2013. Dispersed Personae: Subject-Matters of Scholarly Biography in Nineteenth-Century Oriental Philology. *Asiatische Studien* 67: 1325–1360.

———. Forthcoming. Epistemic Vice: Transgression in the Arabian Travels of Julius Euting. In *Scholarly Personae in the History of Orientalism, 1870–1930,* ed. Christiaan Engberts and Herman Paul. Leiden: Brill.

Vrolijk, Arnoud and Richard van Leeuwen. 2014. *Arabic Studies in the Netherlands: A Short History in Portraits, 1580–1950*. Trans. Alastair Hamilton. Leiden: Brill.

# Weber, Wöhler, and Waitz: Virtue Language in Late Nineteenth-Century Physics, Chemistry, and History

**Herman Paul**

**Abstract** This paper compares the occurrence and use of virtue language among physicists, chemists, and historians in late nineteenth-century Germany, with a special focus on obituaries written for the Göttingen professors Wilhelm Weber, Friedrich Wöhler, and Georg Waitz. Although virtue language was far more prevalent in Waitz's necrologies than in those commemorating Wöhler and Weber, historians, chemists, and physicists resembled each other in that they invoked epistemic virtues if and only if they felt that defining features of what it took to be a scholar were at stake. For all of them, epistemic virtues were shorthand for scholarly personae that they invoked at moments when they perceived those personae as being under pressure. More concretely, categories of virtue and vice served as means for taking sides in debates about such fundamental issues as the proper relation between academy and industry or the relative importance of source criticism in relation to writing – aspects of scientific work that made different demands on scientists in terms of the virtues or dispositions they required.

**Keywords** Epistemic virtues • Scholarly personae • Aims of science • History of physics • History of chemistry • History of historiography • Science and industry • Wilhelm Weber • Friedrich Wöhler • Georg Waitz

## Introduction

If the Göttingen professors Friedrich Wöhler, Wilhelm Weber, and Georg Waitz had one thing in common, it was that they were immortalized in stone by the Berlin sculptor Ferdinand Hartzer. In the years between 1887 and 1899, Hartzer created a bust of Waitz, the historian, a statue of Wöhler, the chemist, and a *Doppelstandbild* depicting Weber, the physicist, together with Carl Friedrich Gauß (Arndt and Arndt

H. Paul (✉)
Institute for History, Leiden University, Leiden, The Netherlands
e-mail: h.j.paul@hum.leidenuniv.nl

 
J. van Dongen, H. Paul (eds.), *Epistemic Virtues in the Sciences and the Humanities*, Boston Studies in the Philosophy and History of Science 321,
https://doi.org/10.1007/978-3-319-48893-6_7

1986: 158–160, 164–166, 188–194).[1] A standard feature of the festive culture of Wilhelmine Germany was that the unveiling of such sculptures was not only accompanied by draperies of flags, festive cantatas, and lavish dinners for invited guests, but also by speeches that described in superlative terms the accomplishments of the men depicted in stone (Arndt 1975). In Weber's case, for instance, Waldemar Voigt delivered a glowing 'consecration speech', as exultant as the marching music played before and afterwards, in which the Göttingen physics professor depicted Gauß and Weber in military language as 'heroes of the spirit … who have fought for our fatherland to become a superpower in the empire of science' (N. N. 1899: 825; cf. Michling 1969). Although Waitz's bust, intended as a present for the occasion of his fiftieth *Doctorjubiläum*, was presented without any ceremony, shortly after the unexpected death of the historian, the lack of unveiling speeches was richly compensated by almost thirty obituaries penned by former colleagues and students (Paul 2017b).

What do such commemorative sources – unveiling speeches and obituaries – reveal about epistemic virtues in late nineteenth-century scholarship?[2] A quick glance at the panegyrics on Wöhler, Weber, and Waitz suffices to observe that quite a few speakers and authors employed virtue language in highlighting the achievements of the deceased. At the unveiling of Wöhler's statue, for example, August Wilhelm von Hofmann praised Wöhler's 'diligence', 'steadfast will power', 'unimpeachable honesty', and 'loyalty to the truth' (*Wahrheitstreue*) (N. N. 1890: 836, 845).[3] Likewise, Waitz's students and sympathizers echoed each other in emphasizing his 'enormous diligence', 'relentless thoroughness', and 'penetrating criticism' (Carstens 1887: 372; Fastenrath 1887: 21; Hagedorn 1886: 147).

What to make of such praise of the dead? As Anna Echterhölter and others have argued, scholarly obituaries constitute a peculiar kind of source material. They should neither be taken at face value, as faithful descriptions of an individual's life, nor rejected out of hand as untrustworthy ('de mortuis nil nisi bonum'). Necrologies are useful sources rather to the extent that they reflect culturally sanctioned standards for praise and blame, thereby articulating socially shared expectations, implicit or explicit codes of conduct, and biographical templates for scholars to conform to. Necrologies, in other words, reveal what scholars are expected to do and, more importantly, how they are expected to live (Echterhölter 2012: 35; Weisz 1988; 14–15; Paul 1980; Outram 1978). Accordingly, the question is not whether Wöhler, Weber, and Waitz were as 'accurate', 'loyal', or 'industrious' as the necrologies portray them, but what the virtues ascribed to them reveal about the normative standards of late nineteenth-century German chemists, physicists, and historians. What sort of virtue catalogues did they invoke? And why and under what circumstances did they speak in terms of virtue about their predecessors?

---

[1] Hartzer's correspondence regarding the Waitz bust can be found in the Göttingen State and University Library, inv. no. 2 Cod. Ms. Philos. 182F. Hartzer: Briefe.

[2] On necrological genres in nineteenth-century Germany, see Bogner 2006: 361–377.

[3] Hofmann's speech was also published in Hofmann 1891: 39–63.

Given that these questions are only beginning to receive attention from historians – following a longer-standing interest of philosophers of science in epistemic virtues, or virtues considered as conducive to the acquisition of knowledge and other epistemic goods (Baehr 2011) – there is not much literature on which to rely (see, however, Tollebeek 2011; Schnicke 2015a; Paul 2012, 2015, 2016). In particular, few attempts have been made so far to explain how and why virtue language was employed in nineteenth-century scholarly discourse. The only explanation that has been put forward so far comes from Jo Tollebeek, who argues that virtue language in the humanities (*Geisteswissenschaften*) around 1900 served the goal of strengthening bonds of unity in emerging scholarly disciplines. In this reading, virtues were markers of shared commitments to shared goals or methods, which in turn contributed to a shared sense of belonging. Virtues, in other words, were short-hand descriptions of an ethos or habitus that scholars had to appropriate in order to be included in scholarly communities committed to 'modern', 'scientific' research (Tollebeek 2015).

Arguably, this explanation accounts for some cases, including the festive celebration of the twentieth-fifth anniversary of Waitz's *historische Übungen* in 1874. At this occasion, epistemic virtues such as 'criticism', 'precision', and 'penetration' were indeed presented as dispositions that no modern, 'scientific' historian could afford to lack (Eskildsen 2015; Schnicke 2015b). However, it is as yet unclear how representative this example is, given that comparative research on commemorative practices, especially across disciplinary divides, has hardly been done, even though the need for such research has been recognized long ago (Abir-Am 1999: 5). For this reason, it seems pertinent to compare the virtue language used by historians around Waitz to language of virtue and vice in two other disciplines: physics and chemistry. Although it is not yet common for historians of the sciences and the humanities to engage in such comparisons across the classic divide between *Geistes-* and *Naturwissenschaften*, I think this is a much-needed step toward an 'integrated history' of the sciences and the humanities such as advocated by Rens Bod (2015) and Lorraine Daston and Glenn W. Most (2015). More importantly, such cross-disciplinary comparisons may help prevent premature extrapolation of discipline- or even time- and place-specific findings to other fields of inquiry.

Concretely, I will therefore examine the occurrence and functions of virtue language in obituaries written for the three men mentioned above: Friedrich Wöhler, Wilhelm Weber, and Georg Waitz. I will base my analysis on what is probably not a complete set of necrologies, but a corpus of source texts that includes as many late nineteenth-century obituaries of the three Göttingen professors as I have been able to find. Although the overwhelming majority of these texts is written by German academics, Waitz and especially Wöhler and Weber were also commemorated in French, Spanish, Italian, British, and American journals. For all these obituaries, I identify their authors and examine the purposes these authors had in honoring the dead as well as the meanings they associated with concepts of virtue and vice.

What emerges is an interpretation that differs from Tollebeek's in so far as it focuses not on unity but on difference. In the necrologies examined for this article, epistemic virtues were primarily invoked in situations of disagreement over 'scholarly personae', that is, over models of what it took to be a scholar in terms of dispositions, attitudes, or character traits (Daston and Sibum 2003; Paul 2014, 2016; Algazi 2016). More often than not, virtues such as 'precision' and 'steadfastness' conveyed commitments to scholarly aims that were regarded as being under pressure. At stake in such cases was not the unity of the discipline, but ways of being a scholar, embodied by exemplary individuals, that adherents felt were challenged by competing personae. Interestingly, whether and to what extent such struggles were fought out in necrological discourse varied from context to context. In Göttingen, the man with the smallest statue turned out to cause the greatest amount of disagreement.

## Georg Waitz

While Georg Waitz's bust was only 76 centimeters high, his fame had been great enough to attract students from all over Europe to his *historische Übungen* (a kind of informal research seminar) (Muhlack 2005; Eskildsen 2015).[4] By the late 1880s, these former students had become a rather diverse lot, in terms of age, nationality, employment, political leaning, and religious affiliation.[5] This diversity is reflected in obituaries that appeared from Riga to Madrid, in different languages, from Protestant, Catholic, and Jewish authors alike. To be sure, all necrologies agreed that Waitz had excelled in such virtues as 'meticulousness', 'carefulness', 'precision', and 'love of truth'.[6] These were virtues characteristic of a philological ethos: a way of doing history that resembled classical philology as practiced by Barthold Georg Niebuhr and his pupils (Kolk 1989). It preferred factual accuracy over hypothetical inference and avoided claiming anything about the past that could not be substantiated with published or unpublished source material. But while all necrologies treated Waitz as a key example of this philological ethos, they disagreed on how praiseworthy this was. Let me offer two examples.

One necrology was written by Heinrich von Sybel, who like Waitz had studied with Leopold von Ranke, the alleged father of modern German historiography, but who had developed himself in such a way as to become almost an antitype of the philological ethos embodied by Waitz (Dotterweich 1978). Indeed, his description of himself as 'four-seventh professor and three-seventh politician' (Varrentrapp 1897: 128) had sufficed for Waitz's students to conceive of Sybel as almost an antipode to their own teacher (von Bezold n.d.: 47), who had increasingly preferred sharp boundaries between history and politics. Sybel, then, used his obituary of Waitz to argue that the historian has three tasks – research, interpretation, and writing – each of which requires particular skills and virtues. As politely as possible,

[4] This section draws and expands on Paul 2017b.

[5] A list of former students can be found in Waitz 1913: 92–100.

[6] These were virtues that Waitz himself had ascribed to his teacher Leopold von Ranke: Waitz 1867: 4.

Sybel continued that Waitz, like all great men, had had a single *forte*, and that his philological virtues clearly illustrated that this *forte* had been research rather than interpretation or writing (Sybel 1886: 484, 485; cf. N. N. 1886: 180). Judging by Sybel's own standards, this implied that Waitz had been one-sided – a judgment that was issued more explicitly and less politely in private correspondence, not only by Sybel, but also by Johann Gustav Droysen, Jacob Burckhardt, and Ernst Bernheim, to mention but a few famous names (see, e.g., Ernst Bernheim to Karl Lamprecht, January 2, 1885).

Some of Waitz's most loyal pupils, such as August Kluckhohn, rushed to his defense and used their obituaries to acquit Waitz from the charge that his philological ethos had caused him to lose himself in details (Kluckhohn 1886: 4042, 4314). Taken at face value, these necrologies seem to indulge in self-gratulation by emphasizing the privilege of having sat at Waitz's feet (Kluckhohn 1886: 4042, 4385; Ermisch 1886: 267; Frensdorff 1885: 9). One could almost gain the impression that these former students were celebrating a shared disciplinary 'we'. However, if we read their pieces against the background of Sybel's necrology and other critical assessments of Waitz's merits, their veiled references to 'the derogatory manner' in which Waitz's achievements had been evaluated (Ermisch 1886: 268) become a key to understanding what their more important purpose was. The primary concern of Waitz's students was to rescue the honor of their former teacher as well as to legitimize their own way of doing history by defending the scholarly persona that Waitz had embodied against its rivals (see esp. Stern 1886: 538, 540; Weiland 1886: 7–8, 11; Wattenbach 1887: 9). Concretely: they clang to virtues of accuracy, precision, and shrewdness in order to dissociate themselves from more politicized and/or aesthetically oriented personae of the sort cherished by Sybel, Droysen, and Burckhardt (Paul 2013).

Something similar applies to Hermann Grauert's more than fifty-page necrology, which presented Waitz as a role model for 'exact critical historical research' (Grauert 1887: 99). Unremarkable as this may sound for a former student of Waitz, it is noteworthy that Grauert issued this praise in the journal of the Görres Society – an association of Catholic scholars more familiar with the persona of the apologetic Catholic writer such as embodied by Johannes Janssen, the polemical defender of Catholic truth, than with the 'impartiality' of a 'scientific' historian (Troxler 2007). Grauert's praise is all the more remarkable because Waitz had had a rather bad name among Catholic historians, because of some notoriously condescending remarks he had made about the 'ultramontane' inclinations of Southern German historiography (Waitz 1846: 533–534). Several Catholic colleagues had returned the compliment and accused Waitz of Protestant bias and ignorance about Catholic history writing (Böhmer 1863: 451; Pastor 1875). So, in this context, praise for Waitz's 'rare talent' amounted to dissociation from prevalent Catholic models, or more precisely, to a difficult attempt at reconciling epistemic virtues of the kind embodied by Waitz with religious virtues associated with Janssen. Unsurprisingly, Grauert's unorthodox attempt at reconciling Protestant criticism with Catholic faith caused not a little nervousness within the Görres Society and something close to a crisis at the association's 1887 annual meeting (Klapczynski 2013: 22–23).

In Waitz's necrologies, in sum, epistemic virtues referred to commitments shared at a much smaller scale than the entire historical discipline. They were invoked, not to unite historians of various stripes, but to defend a scholarly persona or *Gelehrtentypus* (Wittstadt 1996; Maurer 1997) in the context of a politically and religiously divided discipline. Can the same be said about physics and chemistry in late nineteenth-century Germany?

## Scientific Discoveries

The obituaries for Weber and Wöhler as well as the speeches delivered by the unveiling of their statues in Göttingen show a rather different picture. The point is not that virtue language was lacking. To the contrary, Weber was frequently praised for his kindness and sobriety, as well as for his civil courage in joining 'the Göttingen Seven' in challenging King Ernst Augustus's unlawful alteration of the Hanoverian constitution in 1837 – an act of civil disobedience that had temporarily costed him his chair in Göttingen.[7] These were moral and civic virtues, however, without epistemic implications, as illustrated by the fact that they were only invoked in short, personal reminiscences near the end of the obituaries, after far more lengthy discussions of Weber's scholarly merits, in which virtue language was strikingly absent.

Indeed, the main difference between Waitz's necrologies and those of Weber and Wöhler is that the latter focus almost exclusively on scientific results, or on what the Göttingen chemist Hans Hübner, writing about Wöhler, called his *wissenschaftlichen Grossthaten* (Hübner 1882: 1, 9). In loving detail, they dwelled on his 'interesting discoveries' and 'monumental research accomplishments' (N. N. 1890: 839; Hofmann 1882: 3220; F. 1883: xviii). By doing so, they paid far more attention to the outcomes of Wöhler's chemical experiments – to his artificial formation of urea in particular – than to the manner in which 'the dean of German chemists' had conducted his research (N. N. 1882a). Wöhler 'executed', 'isolated', 'improved', 'showed', 'discovered', and 'proved', to quote a few telling verbs from a single paragraph in *The American Journal of Pharmacy* (N. N. 1882c: 592). Despite some occasional remarks on 'a mind fertile in methods, and a judgment seldom at fault', the focus was on 'the fruits of well-directed intellectual labor' (N. N. 1882–1883: 464, 463). Weber, likewise, was primarily remembered for his 'chief contributions to science, those for which he is now best known, and will long continue to be known' (Dolbear 1891–1892: 450). What mattered to his obituary writers were his 'immortal achievements' (Basso 1891–1892: 7), which had provided a 'foundation on which all later researchers' could build (Momber 1892: 146–147).

This focus on discoveries allowed for a historicization, or emphasis on distance in time between the deceased scholars and their obituary writers, for which Waitz's students were hardly prepared. Weber, for instance, was described as 'the last

[7] How the Göttingen Seven were remembered and appropriated across the German political spectrum is the subject of Saage-Maaß 2007: 47–84.

representative of that generation of men of science that cast so much luster on the first half of this century' (Mascart 1891: 109), or even as a man 'who had long been walking among us as a monument of beautiful days past' (Schultz 1891: 3). As *The Popular Science Monthly* put it:

> So retired was Weber's life in his later days that, though his fame had not diminished, the world had almost forgotten that he was still in it; and it is said that when, at the meeting of the German naturalists in Berlin a few years ago, the name of Weber was read in the list of those who had taken part in the first meeting held there in 1828, surprise was expressed at recognizing in their octogenarian friend one who had sat there with Berzelius and Ohm and Heim (N. N. 1892: 553).

Similarly, Hübner stated that much of Wöhler's work already belonged to 'the history book of chemistry' (Hübner 1882: 3). He meant this quite literally, in the sense that chemistry textbooks treated 1828 as the birth year of synthetic urea.[8] This was more than fifty years ago, so that even Wöhler himself was reported to have joked that he belonged to the past (Hübner 1882: 3). Given this historicization, some hermeneutical talent was needed for assessing Wöhler's achievements. In Hofmann's language, one had 'to put oneself back [*zurückversetzen*] into that period' and to 'immerse [*versenken*] oneself in the history of chemistry' (N. N. 1890: 837; Hofmann 1882: 3126).

All this, then, was written from the subject position of a successor rather than that of a pupil. Wöhler's and Weber's obituary writers did not look up at their masters, but backward to the origins of their research paradigms. Weber and Wöhler were not portrayed as men whose epistemic virtues had to be imitated, as in Waitz's case, but as scholars who had 'enriched science with many important facts, many of which have come to serve as points of departure for new, fruitful views' (Voit 1884: 231). What posterity owed to them was therefore not a scholarly persona, but a series of discoveries.

## 'Pure' and 'Applied'

Nonetheless, physics and chemistry around 1890 were as internally divided as the historical discipline, if not in terms of politics and religion, then at least over issues of 'pure' and 'applied' science (Lucier 2012; Gooday 2012). In late nineteenth-century Germany, these contested terms were typically invoked in debates over the merits of academic education in comparison to instruction at technical colleges. They also referred to different conceptions of the relation between research and industry and corresponded to different self-images of the scientist (Harwood 2000). This raises the question whether scholarly personae in physics and chemistry were as contested as they were among historians. Weren't Wöhler's and Weber's students as likely as Waitz's to invoke categories of virtue and vice in defense of their preferred models of being a scholar?

---

[8] Critical assessments of this discovery tale are offered by McKie 1944 and Brooke 1971.

In Weber's case, indeed, various authors tried to claim the deceased as a forerunner of their own 'pure' or 'applied' approach to science. While some biographers invoked Weber's 'absent-mindedness' to emphasize his lack of interest in 'practical applications' (Kohlrausch 1892: 55; Momber 1892: 148),[9] others presented Weber as model of a physicist pursuing laboratory research for, ultimately, the benefit of the public. Central to this discussion was Gauß's and Weber's telegraph of 1833. How 'practical' had Weber's intentions been in connecting his laboratory to Gauß's observatory with a three-kilometer-long wire passing over the tower of St. John's Church?

Writing in *Nature*, the mathematician Carl Runge bluntly declared that 'the enormous practical consequences of telegraphy have nothing to do with the scientific merit of the invention'. Consequently, he considered Weber's 'purely scientific work' as 'more glorious' than the telegraph (Runge 1891). The Strasbourg physicist Friedrich Kohlrausch even went so far as to argue that Gauß and Weber, despite the marble plaque in Göttingen ('first electrical telegraph – Gauss Weber – Easter 1833'), could hardly be considered the inventors of the telegraph, because the men could not even have imagined the practical implications of their finding (Kohlrausch 1892: 54). Felix Koerber, by contrast, emphasized the 'truly practical use' of the telegraph – although it could send no more than six words per minute – and, consequently, Weber's significance 'for the development of electrical telegraphy' (Kbr. 1891: 529). Albert Momber also challenged Runge's interpretation by recalling that Weber had tried to persuade the directorate of the Leipzig-Dresden Railway Company to install a telegraph along their railway – a plan that had eventually been abandoned, but clearly demonstrated Weber's interest in technical applications (Momber 1892: 151; cf. N. N. 1864: 45).

It was in this context of controversy that Weber's epistemic virtues became an argument, especially for scholars committed to 'pure' research. For as John Tyndall, the Irish physicist, had argued under reference to Weber: 'the discoverer and applier of natural truth' require different sets of epistemic virtues, or '[d]ifferent qualities of mind and different habits of thought' (Tyndall 1873: 220). The most explicit reference to such qualities of mind can be found in a necrology by August Voller, a man who, curiously enough, has become best known for the fingers of his right hand. (An X-ray photograph of this hand was included in a widely read report on Wilhelm Röntgen's discovery of X-rays and has been reproduced countless times since then [Glasser 1959: 26–27].) Voller was director of the Physical State Laboratory in Hamburg, where Röntgen conducted his experiments, and a former student of Weber. Like Waitz's pupils, he elaborated at length on his teacher's *wissenschaftlichen Sinn*, 'shrewdness', 'indefatigability', and 'versatility of mind' (Voller 1891: 29). Why did Vollner emphasize those epistemic virtues?[10]

Arguably, Voller found himself torn between, on the one hand, his scientific ambitions and, on the other, the expectations of city magistrates that regarded Voller's State Laboratory as a physical consultancy agency. It was part of the

[9] The historical roots of this *topos* are explored in Algazi 2001, 2005.

[10] A more extensive analysis of the Voller case is provided in Paul 2017a.

director's job, for example, to hold daily office hours for everyone seeking advice in physical matters (Voller 1886: lxiv). In this awkward situation, Voller emphasized again and again that nothing was more important for his lab than to engage in 'serious scholarly work' (Voller 1901: 210–211, 208). Voller found himself, in other words, in an area of tension between 'pure' and 'applied' science, in which he tried to serve the advance of 'pure' science as well as circumstances allowed. Arguably, this was the reason why Voller described Weber's epistemic virtues in glowing terms and why he contrasted his teacher's 'purely scientific goals' with 'the unrest and loud noise' of the 'market of contemporary life' (Voller 1891: 30).

In this case, then, epistemic virtues once again appeared in a context of disagreement about the aims of science, or about the priorities that should be guiding scholarly research. Epistemic virtues were not invoked to confirm shared identities, but to articulate a vision of the 'scientific self' against alternative views of the scholar's true vocation.

## Friedrich Wöhler

If physicists could quarrel over 'pure' versus 'applied' research, so could chemists in late nineteenth-century Germany. As early as the 1840s, Justus von Liebig in Giessen had been criticized in moral terms for exploring the commercial potential of nitrogen-based fertilizer (a true sign of 'greed' and 'vanity', in the eyes of critics more devoted to 'pure' science; see e.g., Mulder 1848: 533–534, 1851: 99). This tension was reflected, moreover, in institutional arrangements. In the days of Wöhler's appointment in Göttingen, the chemistry professor had had to inspect all pharmacies in the Kingdom of Hanover on an annual basis – a requirement clearly reflecting the practical expectations held of the field (Meinel 2002: 100–101, 106–107). Although chemistry in the third quarter of the century had been successful in acquiring more autonomy, the 1870s had witnessed the emergence of new institutional conflicts, primarily between traditional universities devoted to 'pure' research and new, technical colleges founded as training centers for engineers. Surprisingly, however, none of these factions – 'scientists' versus 'industrialists' and 'theoreticians' versus 'factualists' – tried to appropriate Wöhler as an embodiment of their preferred persona. Why not?

Presumably, the answer is that no one cultivated Wöhler's reputation as a father figure as strongly as August Wilhelm von Hofmann, one of the leading German chemists of his day. As a Berlin professor with close ties to the chemical industry (Meinel 1992: 1274), Hofmann saw it as his task to prevent growing tensions between 'science' and 'industry' to deteriorate into conflict. His institutional means to that end was the German Chemical Society, which he had co-founded in 1867 'to provide an opportunity for mutual exchange of ideas between the representatives of speculative and applied chemistry in order to seal anew the alliance between science [*Wissenschaft*] and industry' (N. N. 1868: 3; cf. Johnson 2008: 114–123). Under Hofmann's leadership, this society acquired a reputation for its celebratory style: it let no opportunity pass to celebrate a jubilee or to invite its members to large-scale drinking parties and multi-course dinners (Meinel 1992: 1276–1278). At most of these events, Hofmann delivered festive speeches, in what Friedrich Nietzsche (1995: 83–168) would have called a 'monumental' key, aimed at glorifying the greatness of the scientific ancestors. This construction of a shared past was Hofmann's means for furthering unity among academic and industrial chemists in Germany. If the German Chemical Society at times resembled a Society for Chemical Feasts – the year 1890 even witnessed a *Benzolfest* and an *Anilinfest* (Lepsius 1918: 41–47) – this was because their celebrations of the discipline's past was to provide chemists of different stripes and colors with a shared background. As Christoph Meinel puts it: '[T]he more differentiated and dissonant. …Wilhelmian Germany became, the more brilliant and jubilant were the festivities of the German Chemical Society: expressions of a deep desire to rise above the conflicts of the day, to compensate for dissonance in society with the harmony of a community' (Meinel 1992: 1276).

It is in this context that Hofmann's necrology of Wöhler and his unveiling speech in Göttingen must be read. Both followed well-established patterns, given that the Chemical Society between 1867 and 1918 sponsored no less than twenty-one statues of German chemists, while Hofmann wrote more than fifty lengthy obituaries for the society's journal (Meinel 1992: 1278; Hofmann 1888a). When Hofmann, some nice words on Wöhler's 'lovely character' notwithstanding, portrayed the Göttingen chemist as a pioneer to whose discoveries both academic and industrial chemists were indebted (N. N. 1890: 846, 837–838), this closely resembled his praise for Liebig, back in the 1870s, to whom he claimed the field was indebted for having 'marked out in theory the way of chemical research and discovery for centuries yet to come' (Hofmann 1875: 1069). Likewise, it resembled his obituary of Gustav Magnus, in which Hofmann had highly praised the 'virtues of the *man*', but had strictly separated those from what had taken up the greater part of his necrology: the 'great successes crowning the *scholar's* labors' (Hofmann 1870: 1098, 1097). This focus on Wöhler's historical contribution to the rise of modern chemistry, finally, could become a dominant one, in Germany and abroad, because Hofmann's richly documented piece, complete with long quotations from Wöhler's correspondence (Hofmann 1888b), served as a main source for various other necrologies. While the *Journal of the Chemical Society* simply summarized Hofmann's article, other publications paraphrased it with or without proper source acknowledgment (N. N. 1883b: 258; F. 1883; see also Thorpe 1894: 236–257; N. N. 1893). Wilhelm Ostwald's characterization of Hofmann as 'reigning chemist-in-chief' (Ostwald 1926: 185) therefore also applies to his role in managing Wöhler's memory. It was his example that encouraged chemists in the 1880s to treat Wöhler as an 'origin' rather than as a 'model.'[11]

[11] Hofmann nonetheless allowed himself to describe a difference in working manners between Liebig and Wöhler. Liebig was 'fiery and impetuous, seizing a new thought with enthusiasm, and giving to it the reins of his fancy, tenacious of his convictions, but open to the recognition of error'. In Hofmann's memory, he had been 'irritable and quick to take offence, hot-tempered, hardly

Yet even here, there were two exceptions proving the rule. In the Bavarian Academy of Sciences, Carl von Voit, a former student of Wöhler, emphasized his teacher's 'modesty' and 'simplicity' in order to challenge some prevailing tendencies in his own time: 'Wöhler still was a pure scholar, who valued scholarship above everything …' In apparent contrast to the average chemist in Wilhelmine Germany, Wöhler, said Voit, didn't try to make himself a name with 'fleeting hypotheses' (Voit 1884: 241). This was an appeal to epistemic virtue in the broadest possible sense: a commitment to goods 'internal' to science combined with an ascetic indifference towards such 'external' goods as money and fame – two important *topoi* in nineteenth-century catalogues of vices (Saarloos 2016). Unsurprisingly, this was advocated by a man known as 'a German of the old school', who enjoyed a reputation for his empiricist attitudes and commitment to *Genauigkeit* (precision).[12] Voit, then, emphasized Wöhler's epistemic virtuousness in order to defend a scholarly persona that he perceived as being under pressure, just as *The Chemical News* did when it argued, after an enumeration of Wöhler's 'patience', 'exactitude', 'sobriety of judgment', and 'uprightness': 'Those who maintain that a life devoted to science renders a man harsh, proud, and selfish, will find in the person of Friedrich Wœhler a brilliant refutation of their doctrine' (N. N. 1882b). Again, then, epistemic virtues served as markers of commitments that were seen as in need of defense.

## Conclusion

The 'academic memory cultures' (Zwick 1997; Dicke 2016) surrounding the three men depicted in stone by Ferdinand Hartzer allow for two conclusions regarding epistemic virtues in late nineteenth-century Germany. First of all, I have shown that virtue language was far more prevalent in Waitz's necrologies than in those commemorating Wöhler and Weber. While historians almost without exception commented in detail on especially Waitz's epistemic virtues – that is, on dispositions conducive to such epistemic aims as knowledge and understanding of reality – physicists and chemists reflected on such epistemic virtues only in rare cases. In

master of his emotions'. Wöhler, by contrast, an 'impassioned' man, full of 'equanimity' and 'sobriety', was 'calm and deliberate, entering upon a fresh problem after full reflection, guarding himself against each rash conclusion, and only after the most rigorous testing, by which every chance of error seemed to be excluded, giving expression to his opinion'. This does not imply, though, that Hofmann preferred Wöhler over Liebig. True to his program, Hofmann hastened to declare that both chemists were 'animated by the same intense love of truth', and that their working manners were 'complementary' rather than mutually exclusive. (Hofmann 1882: 3239, here cited in the translation of Thorpe 1894: 255–256). A similar comparison between Liebig and Wöhler was made by Hübner 1882: 10.

[12] Reportedly, Voit had such a dislike of modern academic manners that he could not even bring himself to attending scientific conferences – the gatherings of one local society excepted, because these were social rather than scientific in nature (Frank 1908: xvi, xviii; Lusk 1931: 590; Mitchell 1937: 4, 5).

part, this may be explained by the fact that Weber's and Wöhler's scientific discoveries were a more natural subject for panegyric than Waitz's results: the history of medieval law was a less sensational and less influential field of inquiry than electromagnetism or organic chemistry.[13] Partly also, the different amounts of emphasis that epistemic virtues received in the academic memory cultures around Wöhler, Weber, and Waitz is explained by different styles of commemoration. As I have argued for the case of Wöhler, it was Hofmann's explicit purpose to create a disciplinary past with which both academic and industrial chemists in Imperial Germany could identify. Historians, by contrast, used the occasion of Waitz's death to engage in lively debate over his merits as a role model for younger generations. This might imply that the occurrence of epistemic virtues in scholarly necrologies was at least in part determined by the conventions of academic commemoration in the three respective disciplines. It is well possible indeed that other types of source material – scholarly correspondences, book reviews, or conference reports – will show a greater concern about epistemic virtues among physicists and chemists than is apparent from the necrologies they wrote.

More important, however, for the purpose of this paper is a second conclusion: historians, chemists, and physicists in Imperial Germany resembled each other in that they invoked epistemic virtues when they felt that defining features of what it took to be a scholar were at stake. For all of them, epistemic virtues were shorthand for scholarly personae that they invoked at moments when they perceived those personae as being under pressure. Thus, Waitz's students felt a need to defend their teacher's meticulous source criticism when critics stressed the importance of 'talent for conjecture' and 'literary skill', both of which played only a marginal role in the persona embodied by Waitz. Likewise, Voller emphasized Weber's 'purely scientific goals' over against 'the unrest and loud noise' that he experienced in his own, practically oriented *Staatslabor*, just as Voit idealized Wöhler's devotion to the cause of science in order to justify his criticism of what he perceived as a widespread forsaking of the scholar's true vocation. Epistemic virtues, in short, were invoked in contexts of critical reflection on prevailing scholarly personae.

This, to be sure, does not amount to refutation of Tollebeek's interpretation of virtues as markers of disciplinary identity. It does imply, however, that epistemic virtues served more purposes than enhancing disciplinary unity, not the least of which was critical reflection on modes of being a scholar. Such reflection could take on polemical form, given that scholarly disciplines were no monolithic communities, but fields divided over fundamental issues such as the relative importance of source criticism in relation to *Darstellung*, in the case of history, and the proper relation between academy and industry, in the case of physics and chemistry. Given that such disagreements corresponded to different expectations of the scholar, each with their own specific catalogues of virtue and vice, it is hardly surprising to find scholars quarrelling over scholarly personae and identifying with scholars seen as embodying such personae in exemplary fashion. Therefore, if this chapter has

[13] Sinding (1999: 85) makes a similar argument in order to explain why Louis Pasteur was an 'easier' object of commemoration than Claude Bernard.

shown anything, it is that whenever the Göttingen professors immortalized by Ferdinand Hartzer were hailed for epistemic virtuousness, this happened in defense of particular modes of being a scholar. In late nineteenth-century Germany, epistemic virtues could serve as markers of epistemic commitments and thereby as shorthand for scholarly personae.

**Acknowledgments** I would like to thank Pieter Huistra, Léjon Saarloos, and Axel Wittmann for useful references and Jeroen van Dongen for helpful feedback on a draft version of this chapter. Funding was generously provided by the Netherlands Organization for Scientific Research (NWO).

# Bibliography

## *Primary Sources (Unpublished)*

Bernheim, Ernst to Karl Lamprecht. January 2, 1885. University and State Library Bonn, inv. no. S 2713, Korr. 9.

Hartzer, Ferdinand. 1886–1889. Letters regarding Waitz bust. Göttingen State and University Library, inv. no. 2 Cod. Ms. Philos. 182 F. Hartzer: Briefe.

von Bezold, Friedrich. n.d. Lebenserinnerungen. Bonn University Archive, inv. no. Bh Bez.

## *Primary Sources (Published)*

Basso, Giuseppe. 1891–1892. In commemorazione di Guglielmo Weber. *Atti della R. Accademia delle scienze di Torino: classe di scienze fisiche, matematiche e naturali* 27: 4–13.

Böhmer, Johann Friedrich. 1863. *Briefe*, ed. Johannes Janssen, vol. 1. Freiburg: Herder'sche Verlagsbuchhandlung.

Carstens, C. Er. 1887. Geheimrath Professor Dr. G. Waitz. *Zeitschrift der Gesellschaft für Schleswig-Holstein-Lauenburgische Geschichte* 17: 366–374.

[Dolbear], [Amos E.]. 1891–1892. Wilhelm Eduard Weber. *Proceedings of the American Academy of Arts and Sciences* 27: 449–450.

[Ermisch], [Hubert]. 1886. Georg Waitz † 24. Mai 1886. *Wissenschaftliche Beilage der Leipziger Zeitung*: 265–269.

F., M. 1883. Untitled necrology of Friedrich Wöhler. *Proceedings of the Royal Society of London* 35: xii–xxx.

Fastenrath, Juan. 1887. Los historiadores alemanes Leopoldo de Ranke y Jorge Waitz. *Revista de ciencias históricas* 5: 16–21.

Frank, Otto. 1908. Nachruf Carl Voit gewidmet. *Zeitschrift für Biologie* 51: i-xxiv.

Frensdorff, Ferdinand. 1885. Zur Erinnerung an Georg Waitz: Vortrag auf der Versammlung des Hansischen Geschichtsvereins zu Quedlinburg am 15. Juni 1886 gehalten. *Hansische Geschichtsblätter* 14: 1–10.

Grauert, Herm[ann]. 1887. Georg Waitz. *Historisches Jahrbuch* 8: 48–100.

Hagedorn, A. 1886. Zum Andenken an Georg Waitz. *Mittheilungen des Vereins für Lübeckische Geschichte und Alterthumskunde* 2: 146–149.

Hofmann, Aug. Wilh. 1870. Zur Erinnerung an Gustav Magnus. *Berichte der Deutschen Chemischen Gesellschaft* 3: 993–1098.

———. 1875. The Life-Work of Liebig in Experimental and Philosophic Chemistry; with Allusions to His Influence on the Development of the Collateral Sciences, and of the Useful Arts. *Journal of the Chemical Society* 28: 1065–1140.
———. 1882. Zur Erinnerung an Friedrich Wöhler. *Berichte der Deutschen Chemischen Gesellschaft* 15: 3127–3290.
———. 1888a. *Zur Erinnerung an vorangegangene Freunde: Gesammelte Gedächtnissreden*, 3 vols. Braunschweig: Friedrich Vieweg und Sohn.
———. 1888b. *Aus Justus Liebig's und Friedrich Wöhler's Briefwechsel in den Jahren 1829–1873*, ed. A. W. Hofmann, 2 vols. Braunschweig: Friedrich Vieweg und Sohn.
———. 1891. *Justus von Liebig, Friedrich Wöhler: Zwei Gedächtnissreden*. Leipzig: Veit & Comp.
Hübner, H. 1882. Friedrich Wöhler † (geb. den 31. Juli 1800, gest. den 23. September 1882): Rede gehalten in der öffentlichen Sitzung der K. Gesellschaft der Wissenschaften am 9. December 1882. *Abhandlungen der Königlichen Gesellschaft der Wissenschaften zu Göttingen* 29: 1–11.
K[luckhohn], A[ugust]. 1886. Georg Waitz. *Allgemeine Zeitung*, 4017–4018, 4041–4042, 4313–4314, 4385–4386.
Kbr., F. [= Felix Koerber]. 1891. Wilhelm Weber †. *Himmel und Erde* 3: 528–531.
Kohlrausch, W. 1892. Gedächtnissrede für Wilhelm Weber. In *Jahresbericht des Physikalischen Vereins zu Frankfurt am Main für das Rechnungsjahr 1890–1891*, 43–55. Frankfurt am Main: C. Naumann, 1892.
Lepsius, B. 1918. *Festschrift zur Feier des 50jährigen Bestehens der Deutschen Chemischen Gesellschaft und des 100. Geburtstages ihres Begründers August Wilhelm von Hofmann*. Berlin: R. Friedländer & Sohn.
Lusk, Graham. 1931. Carl von Voit, Master and Friend. *Annals of Medical History* 3: 583–594.
Mascart, [Éleuthère Élie Nicolas]. 1891. Notice sur Wilhelm Weber. *Comptes rendus hebdomadaires des séances de l'Académie des Sciences* 113: 105–109.
Mitchell, H.H. 1937. Carl von Voit. *The Journal of Nutrition* 13: 1–13.
Momber, A. 1892. Zur Erinnerung an Wilhelm Weber: Vortrag in der Sitzung vom 14. Oktober 1891 gehalten. *Schriften der Naturforschenden Gesellschaft in Danzig* 8: 146–154.
Mulder, G.J. 1848. *Scheikundige onderzoekingen, gedaan in het laboratorium der Utrechtsche hoogeschool*. Vol. 4. Rotterdam: H. A. Kramers.
———. 1851. *Scheikundige onderzoekingen, gedaan in het laboratorium der Utrechtsche hoogeschool*. Vol. 5/6. Rotterdam: H. A. Kramers.
N. N. 1864. *Die Leipzig-Dresdner Eisenbahn in den ersten fünfundzwanzig Jahren ihres Bestehens: Denkschrift zur Feier des 7. April 1864*. Leipzig: Giesecke & Devrient.
———. 1868. Constituirende Versammlung vom 11. November 1867. *Berichte der Deutschen Chemischen Gesellschaft* 1: 1–4.
———. 1882a. Utitled death announcement for Friedrich Wöhler. *Journal für Praktische Chemie* 26: 256.
———. 1882b. The Late Professor Wœhler. *The Chemical News* 46: 183.
———. 1882c. Obituary. *The American Journal of Pharmacy* 54: 591–592.
———. 1882–1883. Friedrich Wöhler. Proceedings of the American Academy of Arts and Sciences 18: 463–465.
———. 1883. Friedrich Wöhler. *Journal of the Chemical Society* 43: 258–263.
———. 1886. Leopold v. Ranke und Georg Waitz †. *Zeitung für das höhere Unterrichtswesen Deutschlands* 15: 179–180.
———. 1890. Das Wöhler-Denkmal in Göttingen. *Berichte der Deutschen Chemischen Gesellschaft* 23: 829–851.
———. 1892. Sketch of William Edward Weber. *The Popular Science Monthly* 40: 548–553.
———. 1893. Friedrich Woehler. *The Pharmaceutical Era* 10: 100–101.
———. 1899. Carl Friedrich Gauss und Wilhelm Weber. *Zeitschrift des Vereins Deutscher Ingenieure* 43: 824–825.
Nietzsche, Friedrich. 1995. *Unfashionable Observations*. Trans: Richard T. Gray. Stanford: Stanford University Press.

Ostwald, Wilhelm. 1926. *Lebenslinien: Eine Selbstbiographie*. Vol. 1. Berlin: Klasing & Co..

P[astor], [Ludwig]. 1875. Georg Waitz als preußischer Geschichtsmonopolist. *Der Katholik* 55: 435–445.

Runge, C. 1891. W. E. Weber. *Nature* 44: 272.

Schultz, H. 1891. *Rede am Sarge Wilhelm Webers gehalten am 10. Juli 1891*. Göttingen: Dieterichschen Univ.-Buchdruckerei.

Stern, Alfred. 1886. Georg Waitz. *Die Nation* 3: 538–540.

Sybel, Heinrich v[on]. 1886. Georg Waitz. *Historische Zeitschrift* 56: 482–487.

Thorpe, T.E. 1894. *Essays in Historical Chemistry*. London: Macmillan and Co..

Varrentrapp, Conrad. 1897. Biographische Einleitung. In *Heinrich von Sybel, Vorträge und Abhandlungen*, ed. Conrad Varrentrapp, 1–156. Munich: R. Oldenbourg.

Voit, [Carl] v[on]. 1884. Friedrich Wöhler. In *Sitzungsberichte der mathematisch-physikalischen Classe der k.b. Akademie der Wissenschaften*, 231–242. Munich: F. Straub.

Voller, A. 1886. Physikalisches Staats-Laboratorium. *Jahrbuch der Hamburgischen wissenschaftlichen Anstalten* 3: lxiii–lxv.

———. 1891. Zur Erinnerung an Wilhelm Weber. *Zeitschrift für den Physikalischen und Chemischen Unterricht* 5: 29–30.

———. 1901. Das Physikalische Staats-Laboratorium. In *Hamburg in naturwissenschaftlicher und medizinischer Beziehung: Den Teilnehmern der 73. Versammlung deutscher Naturforscher und Ärzte als Festgabe gewidmet*, ed. [J.] Classen and [Th.] Deneke, 205–212. Hamburg: Leopold Voss, 1901.

Waitz, Eberhard. 1913. *Georg Waitz: Ein Lebens- und Charakterbild zu seinem hundertjährigen Geburtstag 9. Oktober 1913*. Berlin: Weidmannsche Buchhandlung.

Waitz, G. 1846. Deutsche Historiker der Gegenwart: Briefe an den Herausgeber. *Allgemeine Zeitschrift für Geschichte* 5: 520–535.

———. 1867. *Die historischen Übungen zu Göttingen: Glückwunschschreiben an Leopold von Ranke zum Tage der Feier seines fünfzigjährigen Doctorjubiläums, 20. Februar 1867*. Göttingen: W. Fr. Kästner.

Wattenbach, [Wilhelm]. 1887. Gedächtnisrede auf Georg Waitz. In *Akademie der Wissenschaften zu Berlin aus dem Jahre 1886*, ed. Abhandlungen der Königlichen, 1–12. Berlin: Königliche Akademie der Wissenschaften.

Weiland, Ludwig. 1886. *Georg Waitz (geb. 9. October 1813, gest. 24. Mai 1886): Rede gehalten in der öffentlichen Sitzung der K. Gesellschaft der Wissenschaften am 4. Dezember 1886*. Göttingen: Dieterichsche Verlags-Buchhandlung.

## *Secondary Literature*

Abir-Am, Pnina G. 1999. Introduction. *Osiris* 14: 1–33.

Algazi, Gadi. 2001. Gelehrte Zerstreutheit und gelernte Vergesslichkeit: Bemerkungen zu ihrer Rolle in der Herausbildung des Gelehrtenhabitus. In *Der Fehltritt: Vergehen und Versehen in der Vormoderne*, ed. Peter von Moos, 235–250. Cologne: Böhlau.

———. 2005. Geistesabwesentheit: Gelehrte zu Hause um 1500. *Historische Anthropologie* 13: 325–342.

———. 2016. *Exemplum* and *Wundertier*: Three Concepts of the Scholarly Persona. *Low Countries Historical Review* 131 (4): 8–32.

Arndt, Karl. 1975. Denkmäler in Göttingen: Dichter und Gelehrte. In *Göttinger Jahrbuch 1975*, 107–143. Göttingen: Heinz Reise.

Arndt, Monika, and Karl Arndt. 1986. *Der Bildhauer Ferdinand Hartzer 1838–1906*. Berlin: Gebr. Mann.

Baehr, Jason. 2011. *The Inquiring Mind: On Intellectual Virtues and Virtue Epistemology*. Oxford: Oxford University Press.

Bod, Rens. 2015. A Comparative Framework for Studying the Histories of the Humanities and Science. *Isis* 106: 367–377.

Bogner, Ralf Georg. 2006. *Der Autor im Nachruf: Formen und Funktionen der literarischen Memorialkultur von der Reformation bis zum Vormärz*. Tübingen: Max Niemeyer.

Brooke, John Hedley. 1971. Organic Synthesis and the Unification of Chemistry: A Reappraisal. *The British Journal for the History of Science* 5: 363–392.

Daston, Lorraine, and Glenn W. Most. 2015. History of Science and History of Philologies. *Isis* 106: 378–390.

Daston, Lorraine, and H. Otto Sibum. 2003. Introduction: Scientific Personae and Their Histories. *Science in Context* 16: 1–8.

Dicke, Klaus. 2016. Akademische Erinnerungskultur. In *Ambivalente Orte der Erinnerung an deutschen Hochschulen*, ed. Joachim Bauer et al., 21–32. Stuttgart: Franz Steiner.

Dotterweich, Volker. 1978. *Heinrich von Sybel: Geschichtswissenschaft in politischer Absicht (1817–1861)*. Göttingen: Vandenhoeck & Ruprecht.

Echterhölter, Anna. 2012. *Schattengefechte: Genealogische Praktiken in Nachrufen auf Naturwissenschaftler (1710–1860)*. Göttingen: Wallstein.

Eskildsen, Kasper Risbjerg. 2015. Private Übungen und verkörpertes Wissen: Zur Unterrichtspraxis der Geschichtswissenschaft im neunzehnten Jahrhundert. In *Akademische Wissenskulturen: Praktiken des Lehrens und Forschens vom Mittelalter bis zur Moderne*, ed. Martin Kintzinger and Sita Steckel, 143–161. Basel: Schwabe.

Glasser, Otto. 1959. *Wilhelm Conrad Röntgen und die Geschichte der Röntgenstrahlen*. 2nd ed. Berlin: Springer.

Gooday, Graeme. 2012. 'Vague and Artificial': The Historically Elusive Distinction between Pure and Applied Science. *Isis* 103: 546–554.

Harwood, Jonathan. 2000. Das Selbstverständnis des Naturwissenschaftlers im Wandel: Die Lage innerhalb und außerhalb der Akademie zu Beginn des 20. Jahrhunderts. In *Die Preußische Akademie der Wissenschaften zu Berlin, 1919–1945*, ed. Wolfram Fischer, 143–168. Berlin: Akademie Verlag.

Johnson, Jeffrey Allan. 2008. Germany: Discipline, Industry, Profession: German Chemical Organizations, 1867–1914. In *Creating Networks in Chemistry: The Founding and Early History of Chemical Societies in Europe*, ed. Anita Kildebæk and Soňa Štrbáňová, 113–138. Cambridge: Royal Society of Chemistry.

Klapczynski, Gregor. 2013. *Katholischer Historismus? Zum historischen Denken in der deutschsprachigen Kirchengeschichte um 1900: Heinrich Schrörs, Albert Ehrhard, Joseph Schnitzer*. Stuttgart: W. Kohlhammer.

Kolk, Rainer. 1989. Wahrheit, Methode, Charakter: Zur wissenschaftlichen Ethik der Germanistik im 19 Jahrhundert. *Internationales Archiv für Sozialgeschichte der Literatur* 14: 50–73.

Lucier, Paul. 2012. The Origins of Pure and Applied Science in Gilded Age America. *Isis* 103: 527–536.

Maurer, Michael. 1997. Christian Thomasius oder: Vom Wandel des Gelehrtentypus im 18. Jahrhundert. In *Christian Thomasius (1655–1728): Neue Forschungen im Kontext der Frühaufklärung*, ed. Friedrich Vollhardt, 429–444. Tübingen: Niemeyer.

McKie, Douglas. 1944. Wöhler's 'Synthetic' Urea and the Rejection of Vitalism. *Nature* 153: 608–610.

Meinel, Christoph. 1992. August Wilhelm Hofmann: 'Reigning Chemist-in-Chief'. *Angewandte Chemie: International Edition in English* 31: 1265–1282.

———. 2002. Friedrich Wöhler und die Chemie in Göttingen. In *Die Wissenschaften in der Akademie: Vorträge beim Jubiläumskolloquium der Akademie der Wissenschaften zu Göttingen im Juni 2000*, ed. Rudolf Smend and Hans-Heinrich Voigt, 93–109. Göttingen: Vandenhoeck & Ruprecht.

Michling, Horst. 1969. Vom Gauß-Weber-Denkmal und seiner Einweihung. *Mitteilungen der Gauss-Gesellschaft* 6: 16–21.

Muhlack, Ulrich. 2005. Die Stellung von Georg Waitz in der deutschen Geschichtswissenschaft des 19. Jahrhunderts. In *Jaroslav Goll a jeho žáci*, ed. Bohumil Jiroušek, Josef Blüml, and Dagmar Blümlová, 165–181. České Budějovice: Jihočeská univerzita.

Outram, Dorinda. 1978. The Language of Natural Power: The 'Eloges' of Georges Cuvier and the Public Language of Nineteenth Century Science. *History of Science* 16: 153–178.

Paul, Charles B. 1980. *Science and Immortality: The Éloges of the Paris Academy of Sciences (1699–1791)*. Berkeley: University of California Press.

Paul, Herman. 2012. Weak Historicism: On Hierarchies of Intellectual Virtues and Goods. *Journal of the Philosophy of History* 6: 369–388.

———. 2013. The Heroic Study of Records: The Contested Persona of the Archival Historian. *History of the Human Sciences* 26 (4): 67–83.

———. 2014. What Is a Scholarly Persona? Ten Theses on Virtues, Skills, and Desires. *History and Theory* 53: 348–371.

———. 2015. Virtue Language in Nineteenth-Century Orientalism: A Case Study in Historical Epistemology. *Modern Intellectual History*. doi.org/10.1017/S1479244315000293.

———. 2016. The Virtues and Vices of Albert Naudé: Toward a History of Scholarly Personae. *History of Humanities* 1: 327–338.

———. 2017a. What is the Good of Government Interference in Science? A Question from Late Nineteenth-Century Germany. In *The Dutch National Research Agenda in Perspective: A Reflection on Research and Science Policy in Practice*, ed. Beatrice de Graaf, Alexander Rinnooy Kan, and Henk Molenaar, 167–179. Amsterdam: Amsterdam University Press.

———. 2017b. The Virtues of a Good Historian in Early Imperial Germany: Georg Waitz's Contested Example. *Modern Intellectual History*, doi.org/10.1017/S1479244317000142.

Saage-Maaß, Miriam. 2007. *Die Göttinger Sieben: Demokratische Vorkämpfer oder nationale Helden? Zum Verhältnis von Geschichtsschreibung und Erinnerungskultur in der Rezeption des Hannoverschen Verfassungskonfliktes*. Göttingen: V&R Unipress.

Saarloos, Lejón. 2016. Virtue and Vice in Academic Memory: Lord Acton and Charles Oman. *History of Humanities* 1: 339–354.

Schnicke, Falko. 2015a. *Die männliche Disziplin: Zur Vergeschlechtlichung der deutschen Geschichtswissenschaft 1780–1900*. Göttingen: Wallstein.

———. 2015b. Rituale der Verkörperung: Seminarfeste und Jubiläen der Geschichtswissenschaft des 19. Jahrhunderts. *Zeitschrift für Geschichtswissenschaft* 63: 337–358.

Sinding, Christiane. 1999. Claude Bernard and Louis Pasteur: Contrasting Images Through Public Commemorations. *Osiris* 14: 61–85.

Tollebeek, Jo. 2011. *Men of Character: The Emergence of the Modern Humanities*. Wassenaar: Netherlands Institute for Advanced Studies in the Humanities and Social Sciences.

———. 2015. Commemorative Practices in the Humanities around 1900. *Advances in Historical Studies* 4: 216–231.

Troxler, Walter. 2007. *Ein Aussenseiter der Geschichtsschreibung: Johannes Janssen 1829–1891: Studien zu Leben und Werk eines katholischen Historikers*. Berlin: Pro Business.

Weisz, George. 1988. The Self-Made Mandarin: The *Éloges* of the French Academy of Medicine, 1824–47. *History of Science* 26: 13–40.

Wittstadt, Klaus. 1996. Der Enzyklopädist und Polyhistor als neuzeitlicher Gelehrtentypus. In *Literaten, Kleriker, Gelehrte: Zur Geschichte der Gebildeten im vormodernen Europa*, ed. Rudolf W. Keck, Erhard Wiersing, and Klaus Wittstadt, 269–287. Cologne: Böhlau.

Zwick, Jochen. 1997. Akademische Erinnerungskultur, Wissenschaftsgeschichte und Rhetorik im 19. Jahrhundert: Über Emil Du Bois-Reymond als Festredner. *Scientia Poetica* 1: 120–139.

# Virtues of Courage and Virtues of Restraint: Tyndall, Tait and the Use of the Imagination in Late Victorian Science

**Léjon Saarloos**

**Abstract** This paper offers an analysis of the views of two Victorian physicists, John Tyndall and Peter Guthrie Tait, on the scientific use of the imagination. John Tyndall saw the imagination as a way to overcome the limits of experience in science, while Peter Guthrie Tait, by contrast, regarded an unrestrained imagination as a danger to inductive science. At stake in their debate, I will argue, was not just the position of the imagination in science as such, but rather the set of epistemic virtues that both men prescribed in dealing with the imagination. Tait stressed virtues of caution and restraint, where Tyndall urged men of science to practice tolerance and courage. However, historians of epistemic virtue in the nineteenth century have primarily focused on the self-abnegating virtues promoted by Tait, while the virtues championed by Tyndall have often been qualified as primarily moral or aesthetic virtues. Through an analysis of the debate between Tyndall and Tait, this paper argues, first, that accounts of epistemic virtue in the nineteenth century tend to focus too narrowly on self-abnegating virtues of restraint, and secondly, that this can be remedied through a 'weak' reading of the adjective 'epistemic' – one that takes into account the multiple and entangled orientations of virtues and the context in which virtue-language was employed.

**Keywords** Epistemic virtues • Imagination • Courage • Scientific self • History of physics • John Tyndall • Peter Guthrie Tait • Victorian England

## Introduction

At the meeting of the British Association for the Advancement of Science (BAAS) in Liverpool on September 16, 1870, John Tyndall (1820–1893), the eminent Irish physicist and Professor at the Royal Institution in London, treated his audience to an

---

L. Saarloos (✉)
Leiden University, Leiden, The Netherlands
e-mail: j.j.l.saarloos@hum.leidenuniv.nl

 
J. van Dongen, H. Paul (eds.), *Epistemic Virtues in the Sciences and the Humanities*,
Boston Studies in the Philosophy and History of Science 321,
https://doi.org/10.1007/978-3-319-48893-6_8

evening lecture on scientific methodology.[1] Set just after dinner, the lecture was meant to stimulate conversation and therefore had a more popular character than the usual specialist lectures to the various sections of the BAAS. Around 3000 people attended the event, amongst them 'many of the most brilliant ornaments of British science and the representatives of foreign philosophy' (N. N. 1870d).

Indeed, the BAAS meeting in Liverpool that year attracted a large number of eminent men of science from a variety of disciplines, including Peter Guthrie Tait (1831–1901), Francis Galton (1822–1911), James Clerk Maxwell (1831–1879), and the president of the BAAS in that year, Thomas Henry Huxley (1825–1895) (N. N. 1871). Tyndall's after-dinner speech was integrally published and various newspapers and periodicals reported on the event. The lecture thus reached a large audience, consisting both of members of the public and specialist men of science.[2]

If good after-dinner speeches arouse debate and conversation, Tyndall's lecture was superb: it was unmistakably spectacular both in content and style and 'elicited thunders of applause from an audience more than usually critical' (N. N. 1870d). The main subject of the lecture was Tyndall's view on 'the scientific use of the imagination' (Tyndall 1870: 13). In his eyes, the scientific imagination was 'the mightiest instrument of the physical discoverer', and, he continued, men of science should exhibit great feats of 'courage' and 'manful willingness' to make use of this instrument (Tyndall 1870: 16, 44). Tyndall even claimed that 'the thing to be encouraged here is a reverent freedom', rather than caution or restraint (Tyndall 1870: 43). Freedom to use the imagination, Tyndall stated, would enable men of science to see beyond their senses, bring unity to their observations and extend scientific thought beyond the boundaries of Victorian knowledge (Tyndall 1870: 16, 18, 41).

However, this advocacy of a courageous use of the imagination was more contentious than present-day readers might assume (see Brown 2013; Daston 1998). In fact, critics from both within and outside the scientific community attacked Tyndall furiously for the views he expounded. *The Record*'s reviewer dubbed Tyndall's philosophy a 'child of unbridled pride', which appealed only to a 'corrupted human nature' (N. N. 1870b), while a *Times* reviewer, reflecting on the speculative character of Tyndall's lecture, spoke not of the scientific use of the imagination, but of 'the imaginative use of science' (N. N. 1870a). The most trenchant criticism was formulated by the mathematical physicist Peter Guthrie Tait, who attacked Tyndall for his lack of restraint and caution in using the imagination, and deplored the fact that Tyndall's spectacular lecture now legitimized pseudo-scientists to publish their faulty views obtained through their wild imaginations (Tait 1871a: 395).

Interestingly, both Tyndall and Tait, despite their strong differences of opinion, discussed the question of how to use the scientific imagination in terms of virtues, or acquired character traits. Tyndall promoted virtues of courage and manful willingness in the use of the imagination, while Tait urged to practice virtues of restraint and caution. Central to a safe and productive use of the human faculty of

---

[1] For a history and characterization of the British Association for the Advancement of Science, see Morrell and Thackray 1981.

[2] For the term 'man of science', see Baldwin 2015: 4–19.

the imagination, for Tyndall and Tait, was the performance of specific character traits that were conducive to the project of science, although they disagreed about which virtues should be central. The debate between Tyndall and Tait over the imagination did not center on the faculty of the imagination itself, but rather on how it was to be employed, what power it should be given and towards which goals it should be oriented.

Such acquired and truth-conducive character traits, called epistemic virtues in modern scholarship, have attracted much attention in recent years. Both virtue epistemologists and historians of science have used the concept of epistemic virtue to elaborate on the relationship between knowledge and the knower. Philosophers like Linda Zagzebski (1996), Robert C. Roberts and W. Jay Wood (2007), and Jason Baehr (2011) have offered accounts of specific epistemic virtues and their role in ideal scientific inquiry, while historians have explored how character and epistemic virtues functioned in specific contexts in the history of science (for example Goldgar 1995; Levine 2002; Jones 2007; Kivistö 2014).

These historians have shown that epistemic virtues are not only useful concepts for analyzing ideal practices of inquiry, but that they also represent recurring terms in historical sources that address the practice of 'good' scholarship or science, or what was 'bad' science (see Tollebeek 2011; Eskildsen 2013; Kidd 2014). An important observation is that virtues were always oriented towards specific 'goods' (Paul 2014). These 'goods' could be epistemic, political, religious, moral, or aesthetic, and historians have shown that in the nineteenth century, virtues could be oriented towards multiple and entangled 'goods' at the same time, making it hard to narrowly classify specific virtues as purely epistemic, political or moral (Creyghton et al. 2016).

However, recent scholarship has primarily focused on the set of epistemic virtues propagated by Tait: virtues of restraint and caution. These were oriented towards a disciplining of the self in order to achieve objectivity, and stood in a long tradition of asceticism (Levine 2002: esp. chapter 11; Daston and Galison 2007: esp. 203–216; and Bellon 2014). This work has shown that men of science had to actively restrain their subjectivity in order to come to reliable knowledge, leading to the primacy of self-disciplining virtues like patience, industry, or caution; virtues that were oriented towards an inductive, restrained, and 'Newtonian'[3] science (Daston and Galison 2007: 203, 232, 242). In George Levine's words, Victorian scientists were literally 'dying to know': they felt they could only come to knowledge through 'acts of moral self-restraint and refusals of desire' (Levine 2002: 88). Because they were aimed at a practice of science devoid of subjectivity, which cherished inductive reasoning over *a priori* approaches, virtues of restraint and caution have usually been classified as *epistemic* virtues.

The virtues propagated by Tyndall in the above example – courage and willingness – are usually classified as *moral* or *aesthetic* virtues, rather than as epistemic virtues (Daston and Galison 2007, 246–247; Jacobson 2005). These virtues have been described as oriented at an assertion of the self, rather than at self-abnegation,

[3] For the meanings of 'Newtonian' science in Victorian Britain, see Bellon 2014: 222–223.

and they gradually became associated, not with objective science, but with the subjectivity of art (Daston 1998: 82–83). However, the example of Tyndall has also frequently, and convincingly, been employed by historians as a figurehead of objectivity and 'dying to know' epistemology (Levine 2002: 4–5; Stanley 2015: 8; also Kidd in this volume), because he also advocated the performance of virtues of 'self-renunciation' and 'patient industry' (Tyndall 1856: 291). Why, then, did Tyndall champion virtues of courage and willingness in this specific debate over the imagination? And what does this reveal about the ideals of objectivity and 'dying to know' epistemology in the nineteenth century?

The aim of this paper is threefold. First of all, I will analyze the positions taken up by Tyndall and Tait in their debate over the imagination in science by focusing on the virtues they prescribed in using this faculty, and the reasons they had for promoting these virtues specifically. I will first discuss the contents and background of Tyndall's lecture and will then address the two main criticisms forwarded by Tait. Secondly, I will argue that in order to understand this debate, it is necessary to adopt a 'weak' definition of epistemic virtue, that is, one that takes into account the context in which virtue-language was employed and the multiple 'goods' to which a virtue could be oriented (Creyghton et al. 2016). Tyndall advanced his view of the imagination explicitly as an attack on what he saw as a detrimental epistemological tradition, one that Tait defended. A 'weak' definition of epistemic virtue, one that treats its epistemological orientation as one orientation amongst many, makes clear that in this case, courage and willingness – usually categorized as moral or aesthetic virtues – also functioned as epistemic virtues, and were aimed at epistemic 'goods'. Finally, and on a more general level, this paper aims to broaden the narratives of self-annihilation or self-effacing objectivity in the nineteenth century that others have presented, by showing that multiple conceptions of epistemic virtue and virtuous science coexisted and competed, and that the context in which virtue-language was employed is important for our correct reading of these virtues.

## Tyndall and the Scientific Use of the Imagination

To understand the debate on the use of the imagination in late Victorian Britain and the virtues employed in this debate, it is first of all necessary to take a closer look at John Tyndall and the views expressed in his evening-lecture at the BAAS-meeting in Liverpool in 1870. Who was John Tyndall? Tyndall, after studying under Robert Bunsen in Marburg, rose to scientific fame early in the 1850s with his work on diamagnetism. Later in his career, Tyndall's interests broadened to include glaciology, heat, and properties of gases, amongst many other topics.[4] Tyndall has been characterized as an ambitious pioneer of mechanical molecular theory, who aimed 'to arrive at a more definite idea of the molecule itself . . . in a period when there was

[4]For a lengthy obituary notice, in which Tyndall's life and scientific work is described, see F[rankland] 1894.

no consensus among scientists about the specific characteristics of the individual particle' (Yamalidou 1999b; 327). Fellow physicists, Yamalidou continues, were also engaged in theorizing about molecules and particles, but eschewed definite statements about the exact composition of an individual particle (see also Harman 1982: 133–139).

Moreover, Tyndall became known as one of the greatest popularizers of science as Professor of Natural Philosophy at the Royal Institution, lecturing to lay audiences and publishing books for the general public, like his *Fragments of Science for Unscientific People* (Tyndall 1871). Tyndall was part of the London-based X-club, a club of influential British men of science that sought to replace the old clerical scientific establishment with a new, properly scientific, vanguard of scientific naturalists. Other members were for example Thomas Henry Huxley, Edward Frankland (1825–1899), and Herbert Spencer (1820–1903). Members of the X-club sought to transform science, and through science society (Barton 1990; Dawson and Lightman 2014). Tyndall's spot in the limelight and his scientific naturalist ambition frequently pitted him against many laymen, churchmen and fellow men of science. The controversy following Tyndall's take on the imagination at the BAAS in 1870 was one of the many controversies in which he was involved.

Tyndall started his lecture, 'On the Scientific Use of the Imagination', by discussing a suggestive methodological problem considering the unseen phenomena of nature. He asked his audience: 'How then are those hidden things to be revealed? How, for example, are we to lay hold of the physical basis of light, since, like that of life itself, it lies entirely without the domain of the senses?' (Tyndall 1870: 16) The key ingredients of inductive science – experience, observation, and experiment – would not suffice, he claimed. Tyndall stated there was only one solution in overcoming the lack of experience:

> We can also magnify, diminish, qualify, and combine experiences, so as to render them fit for purposes entirely new. We are gifted with the power of Imagination, – combining what the Germans call Anschauungsgabe and Einbildungskraft – and by this power we can lighten the darkness which surrounds the world of senses (Tyndall 1870: 16).

Tyndall sought to overcome the limitations of experience by the use of the imagination. For Tyndall, the imagination had two components: *Anschauungsgabe* and *Einbildungskraft*. Both can be traced back to Kant's epistemology. *Anschauungsgabe*, in Tyndall's view, was the human capability to picture something in the mind that could not be experienced directly, for example the way sound waves travelled through air:

> We travel a very little way from downright sensible experience. Still the imagination is to some extent exercised. The bodily eye ... cannot see the condensations and rarefactions of the waves of sound. We construct them in thought, and we believe as firmly in their existence as in that of the air itself (Tyndall 1870: 18).

This example of the imagination as a way of imaging reality was not very controversial. In fact, without using the concept of imagination, even the mathematical physicists Peter Guthrie Tait and James Clerk Maxwell, who attacked Tyndall after his lecture, drew upon the power to always 'keep the physical entities before his mind' when working with abstract mathematical formulae (Crowe 1967: 134, 213). Moreover, Daniel Brown has argued that this conception of the imagination was broadly shared from the 1860s onward, because new scientific theories, like the science of energy, operated on a level unseen; they had to be constructed in the mind's eye, through 'faculties of mental vision' (Brown 2013: 142–144).

As Tyndall went on, this way of using the imagination to picture physical events was extended, through analogy, from explaining sound waves, to providing explanations of light, color, the atmosphere, the ether, and, finally, hypothesizing on the molecular composition of matter. In Tyndall's hands, the imagination became more than a way of imaging reality; it was to be used as an explanatory and creative force of its own, and this is what Tyndall meant by *Einbildungskraft*. Tyndall envisioned the creative power of the imagination as follows: 'There is in the human intellect a power of expansion – I might almost call it a power of creation – which is brought into play by the simple brooding upon facts' (Tyndall 1870: 18). So, in Tyndall's eyes, a fruitful use of the imagination in science employed both its components: seeing beyond the senses and building and hypothesizing upon these notions.

More controversially, Tyndall contended that real scientific facts and theories could be established by properly employing the imagination (Tyndall 1870; 20, 21, 29). Through *Anschauungsgabe* and *Einbildungskraft*, the man of science could access the unseen world of nature and the facts discovered in this way were 'real' in the sense that they transcended the position of model or hypothesis. This was not straightforward. Although the use of analogy in science was widely practiced, the exact way in which analogies were drawn and what evidence for real physical facts analogies provided was very much open to debate (Kargon 1969). A contrasting position is represented by James Clerk Maxwell, who employed analogies not as a means to establish scientific facts about reality, but as heuristic devices that helped him think through a problem, and even though Maxwell himself used and contributed to molecular theories, he consistently distinguished between mathematical and statistical mechanical models of reality, and the constitution of reality itself (Kargon 1969: 433–434; Stanley 2008; Harman 1998: especially 175–187). Moreover, it was felt by Maxwell, analogies too often proceeded from false assumptions and hypotheses, rather than from experimental results. Instead, he urged to always 'begin with the phenomena' (Maxwell 1890; 309).

Tyndall recognized that the creative role of the imagination was controversial: 'there are tories even in science who regard the imagination as a faculty to be feared and avoided rather than employed. They had observed its action in weak vessels and were unduly impressed by its disasters. But they might with equal justice point to exploded boilers as an argument against the use of steam' (Tyndall 1870: 16). For Tyndall, who presented himself as a progressive, only traditionalist 'tories' would fear the use of the imagination. But, as he then observed, only 'weak vessels' would fail in properly using their faculties, where true men of science would certainly succeed. What, then, made a strong vessel?

In safeguarding a proper use of the imagination, Tyndall called upon a second faculty: 'Bounded and conditioned by cooperant Reason, imagination becomes the mightiest instrument of the physical discoverer' (Tyndall 1870: 16). The role reason had to play in scientific pursuits was to be the conscience of the imagination; it had to check its deductions and hypotheses, but only after the scientific imagination had produced them. This method always landed men of science 'on the solid shores of fact', and a result reached in this way was 'something more than a mere figment of the scientific fancy' (Tyndall 1870: 19–20). In Tyndall's eyes, men of science should use the imagination to come to new results and theories, and then check them by experiment (Tyndall 1870: 19). The imagination, then, was the primary faculty in science and had to be recognized as 'authoritative' (Tyndall 1870: 21). Moreover, only through acts of imagination could a more definite idea of the molecular nature of matter be conceived (Yamalidou 1999a; 233–234). Tyndall's convictions that real physical facts could be established through the imagination, and that the imagination held primacy over reason, stood in marked contrast to Maxwell's distinction between statistical molecular models and reality, and in his view it were experiments that held primacy over the imagination. Maxwell's view of analogy as a heuristic tool, therefore, was much more modest than Tyndall's, who used analogy to establish real physical facts.[5]

## Enabling the Imagination

Well aware of his controversial position, Tyndall eloquently advocated virtues of tolerance for those men of science that did use the imagination properly. Invoking the names of Charles Darwin (1809–1882), Isaac Newton (1643–1727), and William Thomson (1824–1907), Tyndall made a strong point for the imagination, framing these men as exemplars of how a proper use of the imagination led to great results. Thomson was praised for being 'aided by this faculty' and bringing to bear his 'wonderful imagination upon his subject', while Newton was said to make 'leaps of imagination', and 'in the case of Mr. Darwin, observation, imagination, and reason combined have run back with wonderful sagacity and success' (Tyndall 1870: 16, 40, 42). As we shall see, this appeal for tolerance and the appropriation of revolutionary men of science for the cause of the imagination was highly disputed. But Tyndall did so for a reason.

He did so because he was not only trying to make a case for the scientific use of the imagination, but also because he was arguing for a revolutionary kind of science to be pursued. As Maria Yamalidou has observed, Tyndall was aiming to establish 'the idea of material molecularity' through the imagination and he was 'seeking consistently the causal links between observable phenomena and underlying

[5] This, in turn, was related to differences in metaphysical sensitivities and 'deep confidence' in science between Tyndall and the evangelistic Maxwell. Tyndall and the scientific naturalists had 'grand ambitions to know the order of reality', as Ian James Kidd argues in his contribution to this volume (2017), whereas James Clerk Maxwell's confidence in science, was instilled by his deep confidence in the 'close relationship between the physical world and his creator' and his view that the world was ultimately knowable, but that there were 'deeper truths which humans could never know', as Matthew Stanley argues in his chapter in this volume (2017).

mechanisms' (1999a; 234). This project of constructing a molecular mechanical worldview that transcended the position of a model has often been linked to scientific naturalism and the aims of the X-club, who sought to exclude supernaturalism and natural theology from the professional pursuit of science, but also aimed at the popularization of the doctrines of evolution and the atomic theory of matter (Dawson and Lightman 2014; 11). Moreover, scientific naturalism was also a moral project, aimed at 'the bettering of man's estate' as George Levine has observed (2014: 81). These aims became clearer at the end of Tyndall's address. After Tyndall had the imagination explain problems of a growing complexity – climbing from sound waves to light, ether and matter – he asked the ultimate question: where does life come from? This was a question experiments and observation would never be able to answer on their own, Tyndall continued. Luckily, 'the speculative faculty, of which imagination forms so large a part, will nevertheless wander into regions where the hope of certainty would seem to be entirely shut out' (Tyndall 1870: 41). The imagination could boldly go, where no one had dared to go before.

Tyndall presented two controversial options considering the origin of life and related it to the nature of matter. The first he considered presented life as a principle 'inserted into matter at a later date' and the second theory claimed that life 'was present potentially in matter in the nebulous form' (Tyndall 1870: 44). Recognizing the 'absurdity too monstrous to be entertained by any sane mind' (Tyndall 1870: 47) of those options, Tyndall nonetheless advanced his inquiry. He asked why his audience agreed that these options were monstrous and concluded that this was due to the dominant idea of a Cartesian anti-thesis between spirit and matter apparent in British culture.

The reason why his audience feared hypotheses like the origin of life in nebular matter and natural evolution, Tyndall claimed, was historical: 'Supposing that in youth we had been impregnated with the notion of the poet Goethe. .. ; looking at matter, not as brute matter, but as 'the living garment of God'' (Tyndall 1870: 48). Would the scientific community have been brought up in a scientific naturalist fashion, the materialistic idea would have had a greater appeal than the idea of a Cartesian anti-thesis between 'all-vile' matter and 'all-noble' spirit (Tyndall 1870: 48). Tyndall, having dismissed 'all terror and excitement' from his mind, asked his audience not to fear the evolution hypothesis, but to treat it scientifically: 'Under the fierce light of scientific enquiry, this hypothesis is sure to be dissipated if it possess not a core of truth' (Tyndall 1870: 49).

The scientific naturalistic conclusion that all life could be reduced to matter and was once 'latent in a fiery cloud' (Tyndall 1870: 35) was shocking to many, as Tyndall surely recognized. Therefore, he called upon all not to recede into dogmatism or fear, but to act scientifically. In his eyes, 'the thing to be encouraged here is a reverent freedom. .. while the thing to be repressed both in science and out of it, is dogmatism' (Tyndall 1870: 43). Less apologetically, Tyndall attacked all who opposed this freedom and tolerance of the imagination by stating that 'they do not lack the positive element, namely the love of truth, but the negative element, the fear of error, preponderates' (Tyndall 1870: 35). Instead, he stressed that the ideal man of science needed virtues of 'courage' and 'reverence' to overcome error, and a 'desire. .. to know the truth' (Tyndall 1870: 50). To overcome dogmatism and fear, Tyndall's ideal man of science needed 'manful willingness to engage in open contest' (Tyndall 1870: 44).

Concluding these paragraphs on Tyndall, I think three important observations can be made. First of all, Tyndall advocated and popularized his own revolutionary scientific naturalism in an effort to extend the boundaries of scientific inquiry, purge natural theology from the pursuit of knowledge, and to break down what he saw as a bulwark of conservative science, characterized by a fear of hypothesizing about the origin of life and the composition of matter. Secondly, it is worth noting that Tyndall tried to reach these objectives through a lecture on the scientific use of the imagination. Tyndall and his audience recognized that there was a strong connection between the way one practiced science and the 'goods' science was to pursue. This is why Tyndall advocated a free use of the imagination through the exercise of the virtues of courage, manful willingness and tolerance; he believed that these were the character traits that were conducive to 'good' naturalistic science.

Finally, Tyndall juxtaposed the virtues of courage against what he saw as a vicious ascetic tradition of scientific inquiry, one tainted by vices of 'dogmatism' and rendered useless through a 'fear of error', upheld by backward 'tories in science'. In this specific context and with Tyndall's scientific objectives in mind, courage, manful willingness and tolerance functioned as enabling virtues; they were aimed at breaking down an intellectual tradition that cherished fear and dogma over truth and held back the march of scientific naturalism in science. Since Tyndall saw scientific naturalism also as a moral project, the virtues of courage advocated by Tyndall were oriented towards entangled epistemic and moral 'goods'.

Tyndall's plea for men of science to assert themselves, to courageously use the imagination and to tolerate those who do so, seems to contradict the narratives of 'dying to know' and self-restraint for the sake of objectivity as forwarded by Daston and Galison, and Levine, who, as I have stated earlier, even employed the example of Tyndall as a figurehead of 'dying to know' epistemology. However, this is beside the point. In fact, Tyndall sought to remove the obstacles to such a self-disciplined pursuit of knowledge by attacking dogmatic traditions and fears of hypothesizing in science. His plea for these specific virtues was meant to enable men of science to move beyond these restrictive traditions, held by 'tories in science'. True selflessness, the ultimate goal of 'dying to know' epistemology, could only be achieved by abandoning 'all preconceived notions' (Tyndall 1856: 291) in science, and that was the goal Tyndall had set himself in his evening lecture. Taking into account the context in which Tyndall promoted his virtues of courage and tolerance enables us to see that these virtues, usually classified as aesthetic or moral virtues, also fulfilled an epistemic role in this debate. To understand better which tradition of inquiry Tyndall sought to dismantle, the reception of Tyndall's lecture is most insightful, as the defenders of this tradition got a chance to present their arguments.

## Peter Guthrie Tait and Habits of Observation

Tyndall was attacked from various directions on rather different grounds. Prominent were the responses from the clerical side, who criticized Tyndall for his materialistic conclusions. A reviewer of the *English Churchman* wrote: 'What astonishes us beyond measure is, that a man of Professor Tyndall's real ability and earnestness should sneer at the second verse of the Bible … surely, he has here the very thing which he is searching – the true origin of life' (N. N. 1870c). An earlier mentioned reviewer of the *Record,* a Church of England newspaper catering to conservative evangelicals, added that Tyndall's 'God is the God of the Epicureans.… Such a philosophy, the child of unbridled pride of intellect, may appeal to the wildest imagination of corrupted human nature, but it has not sympathy with all the higher yearnings of the soul' (N. N. 1870b). Interestingly, this reviewer explicitly made the connection between Tyndall's suspected atheism and the way he used the imagination.

The most salient scientific response was issued by Peter Guthrie Tait. Unlike Tyndall's clerical critics, Tait did not attack Tyndall's alleged materialism, but rather aimed his arrows at Tyndall's method of using the imagination. In a short, but aggressive review published in *Nature* on March 16, 1871, Tait shredded Tyndall's lecture and the corrupting influence it had on the practice of science:

> Professor Tyndall will eventually have much to answer for. He has lent his authority to the admission of imagination in the pursuit of science, and there is every prospect that people whose imaginative faculty is stronger than their habit of observation will give us all plenty to do. We shall not only have to question nature, but we shall have to eliminate imagination, and thus have two battles to fight for truth (Tait 1871a: 395).

Tait attacked Tyndall because he feared that Tyndall's plea for the use of the imagination in science would lead to disaster, as unworthy men would interpret Tyndall's lecture as an invitation to freely imagine their way into science. Tyndall's conception of the ideal man of science, who would use the primary faculty of the imagination courageously and freely, was an abomination to Tait, who stressed that habits of observation were more important. Although Tyndall had also advocated the use of reason and experiment as a check to imagination, Tait, echoing Maxwell's concerns about the use of analogy in science, militated against this primacy of imagination and hypothesis over habits of observation. Tait's view on the imagination in science, then, stood in marked contrast to Tyndall's and was embedded in a different scientific outlook.

Who was Peter Guthrie Tait, why did he criticize Tyndall, and to what 'goods' was he committed? Tait was Professor of Natural Philosophy at the University of Edinburgh and was primarily occupied with the study of thermodynamics and the new science of energy. Tait had worked together intensively with William Thomson and was a close friend of James Clerk Maxwell, with whom he had studied in Edinburgh and Cambridge. Their mathematical training in the Tripos was recognizable in their work; both Tait's and Maxwell's physical work made use of complex mathematical reasoning.

One reason why Tait entered the fray after Tyndall's lecture was because he and Tyndall represented two different groups of physicists in late Victorian Britain, who competed for scientific authority. Tyndall's group, as we have seen, was closely associated with the scientific naturalistic X-Club and was committed to atomic theories of matter, evolutionary biology, and shared an antipathy to natural theology (Dawson and Lightman 2014; 11). Tait's group, amongst which were James Joule (1818–1889), James Clerk Maxwell, and William Thomson, by contrast, was based in the Northern cities of Manchester, Edinburgh and Glasgow, and was not so much committed to reform, evolution or naturalism, but rather expounded a new science of energy, based on thorough experimentation and rigid mathematics (Smith 1998: 170–191). Tait and Thomson's *Treatise of Natural Philosophy* (1867) and Maxwell's *Treatise on Electricity and Magnetism* (1873) can be seen as the 'most enduring embodiments of the science of energy' (Smith 1998: 288). Moreover, as P. M. Harman has argued, Thomson, Tait, and Maxwell also had strong theological commitments; their science of energy left room for divine agency and design, and was a 'manifestation of the continued influence of the 'natural theology' tradition' (1982; 69) that the group around Tyndall sought to dismantle.

However, Tyndall, on several occasions and also in his lecture on the use of the imagination, appropriated the findings of this new science of energy, combined these insights with the theory of evolution, and portrayed Thomson as a hero for his fruitful use of the imagination (Tyndall 1870: 12–13, 16, 40). Moreover, the naturalistic conclusions about the molecular nature of matter and the origin of life arrived at by Tyndall's use of the imagination were at odds with the religious beliefs of both Maxwell and Tait (Knott 1911: 290–293; Stanley 2008: 52). The North British group felt compelled to respond to this attack on their religious orientations and their ownership of the science of energy, and Tait was the first to deliver a blow in his review of Tyndall's lecture.

As the quote earlier shows, Tait attacked both Tyndall's advocacy of imagination over experiment, and the negative influence this would have on other men of science (Tait 1871a: 395). He offered an example of this bad influence as well: a certain Dr. Cochrane, who in 1870 had published a scientific article on yellow fever in the *Mobile Daily Register*, in which he forwarded an analysis of the disease 'from the imaginative side', as Tait claimed (Tait 1871a: 395). Cochrane imagined yellow fever to be composed of germs and connected 'his speculations into these matters with similar speculations' on other diseases, 'without paying any attention to facts regarding yellow fever' and even stated that he based himself only 'on what he believes, but does not know' (Tait 1871a: 395). For Tait, both Cochrane's conclusions and the way they were reached were an example of an improper use of the imagination in science. Cochrane's approach denied or ignored all progress science had made: 'Are we to live, scientifically, in the same way as alchemists and astrologers did in the Middle Ages? And are we to ignore all that Bacon and Newton have done for us?' (Tait 1871a: 395). Inductive method, responsible for centuries of scientific progress, should not have to give way to imaginative hypothesizing. Tait concluded with an insightful advice: 'Let us use the imaginative faculty by all means; but, in doing so, let us take our stand on the firm ground of the known before we venture ourselves into the unknown.' (Tait 1871a: 395).

Through Cochrane's example, Tait attacked Tyndall and his advocacy of the use of the imagination. Tyndall's view, which stressed tolerance, willingness and courage in the use of the imagination, was starkly contrasted by Tait, who called for 'rigid observation and induction', 'habits of observation', and attributed primacy to the 'firm ground of the known'. Nonetheless, Tait did concede that the imagination was a faculty that had to be used in science. The debate between Tyndall and Tait over the imagination did not center, therefore, on the faculty of the imagination itself, but rather on how it was to be employed, what power it should be given and towards which 'goods' it should be oriented.

How then, did Tait think that men of science should use their speculative faculty? As we have seen, Tait criticized Tyndall on two grounds. Firstly, he attacked Tyndall for prioritizing the imagination over 'habits of observation' and, secondly, he stated that Tyndall failed to base himself on the 'firm ground of the known'. What did Tait mean by that? A source that gives much insight into Tait's conception of what good science was, is the *Treatise of Natural Philosophy* (1867), written collaboratively by William Thomson and Tait, and often referred to as *T&T′*. As Crosbie Smith (1998: 196–199) has observed, *T&T′* was one of the written embodiments of the North-British science of energy and sought to claim a 'Newtonian pedigree' by offering a re-reading of Newton. The authors redefined concepts like 'force', 'work' and 'energy', and in this way sought to base their theory on a solid, accurate and measurable footing (Smith 1998: 201–202). Claiming a Newtonian pedigree also had a more implicit moral dimension. *T&*T′ also appealed strongly to moral language: "accuracy' went hand in hand with such *moral* epithets as 'integrity', 'honesty' and 'commercial fairplay" (Smith 1998: 201). Moreover, as Richard Bellon has shown, the image of Newton in Victorian Britain became shorthand for the inductive method and its guiding virtues of patience, humility, and self-discipline (2014; 222–224). For these reasons, *T&T′* serves as an excellent source from which to analyze Tait's view of the ideal scientific man, the virtues he should cultivate, and the 'goods' to which he was ideally committed.

Considering 'habits of observation', the first component of Tait's criticism of Tyndall, the chapter in *T&T′* on 'Experience' offers a peak into the moral dimension of the new science of energy. The chapter started out by championing experiments as the only method to come to insight into nature, enabling the man of science to study its parts in isolation, in contrast to 'hasty philosophers' that rushed to grand conclusions (Thomson and Tait 1867: 442). Instead, Tait and Thomson argued for 'endless patience and perseverance in designing and trying different methods' and stated that scientific success could only be reached by a person who was 'not allowing himself to be disheartened by the non-success of one form' and 'judiciously varies his methods' (Thomson and Tait 1867: 443). Nature should be 'interrogated' and be handled with 'utmost suspicion', metaphors again pointing to the moral nature of scientific practice (Thomson and Tait 1867: 443–444).[6] Many pages

[6]Although Thomson and Tait did not refer explicitly to Bacon, Evelyn Fox Keller (1985: 33–42) points out that the metaphor of 'interrogation' drew on a masculine Baconian discourse about nature, according to which nature's feminine subjectivity should be subdued.

emphasize that this suspicious interrogation should be done through repeated experiments and 'careful observation' so that 'inaccuracies' could be avoided (Thomson and Tait 1867: 445–447). Furthermore, the step from experimental results to mathematical formulae was discussed in such a way that precision and accuracy was maintained (Thomson and Tait 1867: 454–456). In the following chapter, the practical use of instruments and the best way to obtain accurate measurements was discussed. Again, virtues of 'accuracy', 'carefulness' and 'delicacy' (Thomson and Tait 1867: 462, 464, 470, 472) recur over and over again and were seen to safeguard a correct use of scientific instruments (see also Gooday 2004). Concluding then, when Tait, in his reply to Tyndall, mentioned habits of observation as being more important than the imagination, he referred to the habitual and continuous practice of the epistemic virtues of patience, perseverance, suspicious judiciousness, carefulness, and accuracy, as presented in *T&T'*. The performance of this specific set of virtues is commonly associated with self-disciplined inductive science in the accounts of Levine (2002), Daston and Galison (2007), and Bellon (2014), although singular virtues, like 'perseverance', could have roots in multiple conceptions of science (Stanley 2017: 16–17).

This constellation of virtues conducive to inductive science did not appear just in *T&T'*. In fact, Tait expressed virtues of accuracy, precision and carefulness in many publications – in textbooks, letters, and in critical articles of others' work[7] – and was himself also characterized by others as being very accurate and careful (Knott 1911). Moreover, the many controversies in which Tait was engaged – for example with X-club member Herbert Spencer, or metaphysician and Shakespeare scholar Clement Ingleby – seemed to center around questions of *a priori* reasoning and the value of inductive science versus more 'creative' and 'spurious' science (Tait 1871c).

An earlier controversy between Tyndall and Tait confirms this view. A few years before Tyndall's BAAS lecture, Tait had clashed heavily with Tyndall on the topic of what good science was in a debate about the contributions of Julius von Mayer (1814–1878) to the history of the doctrine of conservation of energy. This debate, fought between 1862 and 1864, was instigated by Tyndall, who claimed that Mayer was the originator of the law of energy conservation, instead of James Prescott Joule, who was generally regarded as the one who had established this principle experimentally (Tyndall 1862; see also Cahan 2012). Although both Thomson and Joule himself responded to Tyndall's attempt to rewrite the history of this law, it was Tait who threw himself into controversy most aggressively. Tait attacked Mayer's claim to authorship of the law of conversation of energy by accusing him of using 'false analogies' and 'speculations' in his rather philosophical work on the mechanical equivalent of heat, and for basing himself on wrong hypotheses, whereas Joule 'saw that experiment, not hypothesis, was required to arrive at knowledge' (Tait 1863; 430). Although Mayer's paper preceded Joule's, Tait argued that the method

[7] See for example Tait's review of Alexander Bain's *Logic*, in which Tait expressed his disgust with Bain's 'looseness of phrase' and warned against Bain's 'dangerous misconceptions of physical truths', Tait 1870; and Maxwell's review of 'Little T&T'' in 1873.

Mayer used disqualified his results. Again, we can see Tait sharply juxtaposing the experimental, inductive method and the mathematical formulation of fundamental laws, to the liberal use of hypotheses and analogies in science (Harman 1982; 62–64; and Lloyd 1970), in an effort to protect the British science of energy from incursions from scientific naturalists like Tyndall.

## Sensation and Science

The second component of Tait's criticism of Tyndall's use of the imagination was aimed at the popular and speculative character of Tyndall's lecture. Tyndall, argued Tait, failed to base himself on the 'firm ground of the known'. Instead, Tyndall seemed to take a 'road through the air' towards knowledge (Tait 1871a: 395). Tait's characterization of Tyndall's 'road through the air', as Daniel Brown has observed, was aimed at the popular character of Tyndall's lecture and the lack of rigid mathematics in Tyndall's popular work. According to Brown, this was 'the crux of North British opposition to Tyndall: he is not a wrangler' (2008: 115). Maxwell, Thomson, and Tait were all wranglers. For the North-British group, mathematical language was the most clear and best resembled reality, whereas Tyndall relied on illustrative experiments during spectacular lectures.[8]

In several articles in *Nature,* Tait had attacked such popular lectures. Considering Rev. Prof. Haughton's scientific lectures on the principle of least action at the Royal Institution, the same institute where Tyndall was professor, Tait lashed out:

> The morbid craving for excitement, which is characteristic of mental indolence, as well as of effete civilization, has led to the introduction of Sensation. .. not merely into our newspapers and novels, but also into our pulpits. It could not be expected that our popular scientific lectures would long escape the contamination. (Tait 1871b: 177).

Tait subsequently blamed Haughton's account for being 'loose and slipshod', because he abandoned 'strictly mathematical applications' and plunged 'headlong into a wild sea of speculation without previous careful definition of his terms' (Tait 1871b: 177). As Tait argued, the lack of strict mathematics and careful language signified 'that we are dealing with Sensation where we looked for Science' (Tait 1871b: 178). Striking a similar note in a review of a book on the nature and origin of comets by the Leipzig professor J. K. F. Zöllner, Tait argued that Zöllner's use of mathematics was at best '"second-year" mathematics (and not always correct)' (Tait 1872: 177). Moreover, since Zöllner ventured out to discredit the findings of British physicists like Thomson and Tait himself, Tait objected to 'the sensational imputation of discourtesy' (Tait 1872: 178). Although the subjects covered by Haughton and Zöllner differed from the work Tyndall presented in his lecture on the imagination, it must be observed that Tait criticized not the scientific content of their work per se, but rather aimed at the carelessness, looseness and inaccuracy of language

[8] On the merits of scientific sensation during the nineteenth century, see Rhys Morus (2011).

and mathematics employed in their work, words he had also reserved for the likes of the logician Alexander Bain, with a citation of whom Tyndall had started his BAAS address in order to legitimize his liberal use of analogy and deduction (Tait 1870; 89–90).

The importance of mathematics and accurate language to safeguard against popular and sensational 'nonsense' was echoed in Maxwell's poems on his friend Tait. Reflecting on Tait's BAAS lecture in 1876, in which Tait made an argument for definiteness in language and mathematics, Maxwell wrote the following lines:

> For Tait comes with his plummet and his line,
> Quick to detect your
> Old bosh new dressed in what you call a fine
> Popular lecture
> …
> But see! Tait writes in lucid symbols clear
> One small equation;
> And Force becomes of Energy a mere
> Space-Variation[9]

Clearly aimed at Tyndall's popular and sensational lectures, this poem asserted Tait's supremacy. Through careful mathematics and rigid reasoning, at least according to Maxwell, Tait had shown in his lecture that the concept of force was to be distinguished from the more fundamental concept of energy (Tait 1876: 459–463). This was an important point, and closely related to the importance of accuracy and definiteness of language and mathematics earlier identified, because many of the scientific men engaged in the science of energy – Tait, Thomson, Rankine, and Maxwell – sought 'to expurgate the doctrine of the conversion of forces from physical theory, distinguishing the energy principle from the concept of force' (Harman 1982; 60), whereas men like Tyndall and Spencer conflated these terms and used them indiscriminately, also in popular lectures and publications (Smith 1998; 185–187).

Concluding then, Tait's conception of the natural scientist contrasted strongly with that of Tyndall. Tait warned for the imagination and urged men of science to employ 'habits of observation', with which he meant that the physical scientist should practice virtues of patience, accuracy, carefulness, and judiciousness; virtues aimed at a self-disciplined pursuit of knowledge. Additionally, when Tait criticized Tyndall for authorizing all kinds of pseudo-scientists to engage in speculation, he did so on the basis of a 'firm ground of the known.' For Tait, this firm ground was held by the combination of careful experiment, rigid mathematics and accuracy of language, which he contrasted to Tyndall's advocacy of the primacy of imaginative hypothesizing over experiment. Moreover, views expounded by sensational lecturers were seen to be loose and slipshod, while Tait's own work excelled in definiteness, precision and carefulness.

---

[9]The full poem by Maxwell is cited in Brown 2008: 118–121. A plummet and line were devices used to measure depth at sea, but serve as a metaphor here to test the intellectual depth of a popular lecture.

Although Tait conceded that the faculty of the imagination should be used in science, he gave primacy to virtues that could restrain the imagination. Tait certainly disapproved of Tyndall's conception of *Einbildungskraft* – the imagination as the most important force in forming new conceptions of reality – but he supported the ideal that men of science should always keep physical entities in their minds when reasoning and mathematizing, which Tyndall referred to as *Anschauungsgabe*. Moreover, just as Tyndall's advocacy of courage in using the imagination was oriented towards scientific naturalism, Tait's plea for a restrained imagination was aimed at the inductive science of energy and was still very much influenced by natural theology. So although these virtues of restraint appear to be epistemic in the strong sense, they also had strong moral and religious layers of meaning. Tait forwarded them as a defense against the religious implications of scientific naturalism and Tyndall's search for the origin of life and matter. As was the case with Tyndall, the virtues propagated by Tait were characterized by an entanglement of both epistemic and moral orientations.

## Conclusion

This chapter has presented two contrasting views of the imagination in late Victorian Britain. For John Tyndall, the faculty of the imagination was the 'mightiest instrument of the physical discoverer'. Not only did the imagination enable scientific men to imagine physical processes that could not be experienced directly through observation, it also served an important function in scientific practice. It could extend the boundaries of knowledge and could lead, through hypotheses from which results could be deduced, to new, and distinctively real, scientific facts that could not be reached otherwise. The faculty of reason would check the hypotheses reached through the imagination, although the latter was regarded as primary. Recognizing the controversial position of this view, Tyndall maintained that the use of the imagination should be directed by virtues of courage, tolerance and manful willingness. These virtues guaranteed a fruitful employment of the faculty of the imagination and safeguarded against a harmfully restrictive tradition, characterized by dogmatism, fear and restraint. Tyndall saw these virtues of freedom as enablers: they made sure that the faculty of the imagination could be properly used in science. Ultimately, these virtues were directed towards what Tyndall saw as 'good science': the project of scientific naturalism.

This view was contrasted by Peter Guthrie Tait, who stood firm to defend the tradition of inductive science. He argued that the habitual performance of virtues like accuracy, restraint and patience should hold primacy over the exercise of tolerance or courage. These virtues of restraint did not enable the imagination, but rather served to discipline this faculty. They served as guardsmen that stopped the advances of faulty analogies and materialistic hypotheses into the Newtonian bastion of inductive science. As was the case with Tyndall, Tait's plea was oriented towards his conception of what 'good' science was: the inductive science of energy.

Much of the literature on epistemic virtues in the late nineteenth century has stressed the tradition of inductive science as personified by Peter Guthrie Tait in this example. Representatives of this tradition stressed ascetic epistemic virtues of restraint, virtues that were to discipline one's self and that were to prevent the self from pressing its subjective stamp on objective science. The catalogue of virtues cherished by Tait had a direct bearing on scientific practices and the results obtained thereby. They were virtues aimed at gaining trustworthy knowledge about nature and have therefore been classified as epistemic virtues. However, Tait's defense of inductive science also had strong moral connotations; Tait, like James Clerk Maxwell, sought to defend his religious worldview against scientific naturalism. The epistemic and moral connotations of these virtues of restraint, then, were entangled and should not be seen apart. At the same time, I have also argued that Tyndall's enabling virtues of courage, manful willingness and tolerance, although they have usually been classified as moral or aesthetic, had a strong epistemic orientation as well. Scientific practitioners, in Tyndall's eyes, needed these enabling virtues to break free from the constraints of an ascetic tradition of scientific practice that he saw as an impediment to the progress of true scientific naturalistic science.

This chapter, finally, has shown that the characterizations of Daston and Galison, and Levine of the late nineteenth century as a period of 'objectivity' and 'dying to know' could be experienced in different ways by individual scientific practitioners. Both Tait and Tyndall sought to decrease the stamp of scientific selves on science, but by different means; Tait through self-discipline and the rigidity of mathematics, Tyndall through the dismantling of a restrictive epistemological tradition. However, both Tyndall and Tait also saw science as a moral project, oriented towards moral 'goods'. A narrow reading of epistemic virtue, one that only pays attention to epistemic 'goods', might obscure the layered nature of these virtues; they were oriented towards multiple and entangled 'goods' at the same time. To conclude this chapter, then, I would like to suggest that students of epistemic virtues in history approach virtue-language through its relation with the multiple 'goods' at which these virtues were aimed, and the context in which they were employed, because only such contextualized accounts of epistemic virtue can show the multiple layers of meaning involved in virtue-language and the multiple stakes involved in scientific debates.

**Acknowledgment** Funding was generously provided by the Netherlands Organization for Scientific Research (NWO).

## Bibliography

### *Primary Sources (Published)*

F[rankland], E[dward]. 1894. John Tyndall. *Proceedings of the Royal Society of London* 55: xvii–xxxiv.

Knott, Cargill Gilston. 1911. *Life and Scientific Work of Peter Guthrie Tait*. Cambridge: Cambridge University Press.

Maxwell, James Clerk. 1873. Thomson and Tait's Natural Philosophy. *Nature* 7 (178): 399–400.
———. 1890. The Scientific Papers of James Clerk Maxwell. *Volume II.* New York: Dover Publications, Inc.
N. N. 1870a. Review of Tyndall's Lecture 'On the Scientific Use of the Imagination'. *Times* (September 19).
———. 1870b. Review of Tyndall's Lecture 'On the Scientific Use of the Imagination'. *Record* (September 23).
———. 1870c. Review of Tyndall's Lecture 'On the Scientific Use of the Imagination'. *English Churchman* (September 29).
———. 1870d. Review of Tyndall's Lecture 'On the Scientific Use of the Imagination'. *Times* (October 3).
———. 1871. *Report of the Fortieth Meeting of the British Association for the Advancement of Science held at Liverpool in September 1870.* London: John Murray.
Tait, Peter Guthrie. 1863. On the Conservation of Energy. *The London. Edinburgh and Dublin Philosophical Magazine and Journal of Science* 25 (170): 429–431.
———. 1870. Energy and Prof. *Bain's Logic. Nature* 3 (57): 89–90.
———. 1871a. Imagination in Science. *Nature* 3 (72): 395.
———. 1871b. Sensation and Science. *Nature* 4 (88): 177–178.
———. 1871c. True and Spurious Metaphysics. *Nature* 5 (109): 81.
———. 1872. Science and Sensation II. *Nature* 6 (140): 177–178.
———. 1876. *Force. Nature* 14 (360): 459–463.
Thomson, William, and Peter Guthrie Tait. 1867. *Treatise of Natural Philosophy.* Oxford: Oxford University Press.
Tyndall, John. 1856. On the Study of Physics. In 1871 published in *Fragments of Science for Unscientific People. A Series of Detached Essays, Lectures, and Reviews,* ed. John Tyndall, 297–320. New York: D. Appleton and Company.
———. 1862. On Force. *The London, Edinburgh and Dublin Philosophical Magazine and Journal of Science* 24: 57–66.
———. 1870. Scientific Use of the Imagination. In *Essays on the Use and Limit of the Imagination in Science*, ed. John Tyndall, 13–51. London: Longmans, Green, and Co.
———. 1871. *Fragments of Science for Unscientific People. A Series of Detached Essays, Lectures, and Reviews.* New York: D. Appleton and Company.

## Secondary Literature

Baehr, Jason. 2011. *The Inquiring Mind: On Intellectual Virtues and Virtue Epistemology.* Oxford: Oxford University Press.
Baldwin, Melinda. 2015. *Making 'Nature': The History of a Scientific Journal.* Chicago: University of Chicago Press.
Barton, Ruth. 1990. 'An Influential Set of Chaps': The X-Club and Royal Society Politics, 1864–85. *British Journal for the History of Science* 23: 53–81.
Bellon, Richard. 2014. There Is Grandeur in This View of Newton: Charles Darwin, Isaac Newton and Victorian Conceptions of Scientific Virtue. *Endeavour* 38: 222–234.
Brown, Daniel. 2013. *The Poetry of Victorian Scientists. Style, Science and Nonsense.* Cambridge: Cambridge University Press.
Cahan, David. 2012. The Awarding of the Copley Medal and the 'Discovery' of the Law of Conservation of Energy: Joule, Mayer and Helmholtz Revisited. *Notes and Records of the Royal Society of London* 66 (2): 125–139.

Creyghton, Camille, Pieter Huistra, Sarah Keymeulen, and Herman Paul. 2016. Virtue Language in Historical Scholarship: The Cases of Georg Waitz, Gabriel Monod and Henri Pirenne. *History of European Ideas* 42: 924–936.
Crowe, Michael. 1967. *A History of Vector Analysis: The Evolution of the Idea of a Vectorial System*. Notre Dame: University of Notre Dame Press.
Daston, Lorraine. 1998. Fear and Loathing of the Imagination in Science. *Deadalus* 127: 73–95.
Daston, Lorraine, and Peter Galison. 2007. *Objectivity*. New York: Zone Books.
Dawson, Gowan, and Bernard Lightman. 2014. Introduction. In *Victorian Scientific Naturalism: Community, Identity, Continuity*, ed. Gowan Dawson and Bernard Lightman, 1–24. Chicago: University of Chicago Press.
Echterhölter, Anna. 2012. *Schattengefechte: Genealogische Praktiken in Nachrufen auf Naturwissenschaftler (1710–1860)*. Göttingen: Wallstein.
Eskildsen, Kasper Risbjerg. 2013. Inventing the Archive: Testimony and Virtue in Modern Historiography. *History of the Human Sciences* 26 (4): 8–26.
Goldgar, Anne. 1995. *Impolite Learning. Conduct and Community in the Republic of Letters*, 1680–1750. New Haven: Yale University Press.
Gooday, Graeme. 2004. *The Morals of Measurement: Accuracy, Irony, and Trust in Late Victorian Electrical Practice*. Cambridge: Cambridge University Press.
Harman, P.M. 1982. *Energy, Force, and Matter: The Conceptual Development of Nineteenth-Century Physics*. Cambridge: Cambridge University Press.
———. 1998. *The Natural Philosophy of James Clerk Maxwell*. Cambridge: Cambridge University Press.
Jacobson, Daniel. 2005. Seeing by Feeling : Virtues, Skills, and Moral Perception. *Ethical Theory and Moral Practice* 8 (4): 387–409.
Jones, Stuart. 2007. *Intellect and Character in Victorian Britain: Mark Pattison and the Invention of the Don*. Cambridge: Cambridge University Press.
Kargon, Robert. 1969. Model and Analogy in Victorian Science: Maxwell's Critique of the French Physicists. *Journal of the History of Ideas* 30 (3): 423–426.
Keller, Evelyn Fox. 1985. *Reflections on Gender and Science*. New Haven, CT: Yale University Press.
Kidd, Ian James. 2014. Was Sir William Crookes Epistemically Virtuous? *Studies in History and Philosophy of Science of the Biological and Biomedical Sciences* 48: 67–74.
———. 2017. *Confidence, Humility, and Hubris in Victorian Scientific Naturalism*. In this volume.
Kivistö, Sari. 2014. *The Vices of Learning: Morality and Knowledge at Early Modern Universities*. Leiden: Brill.
Levine, George. 2002. *Dying to Know: Scientific Epistemology and Narrative in Victorian England*. Chicago: University of Chicago Press.
———. 2014. Paradox: The Art of Scientific Naturalism. In *Victorian Scientific Naturalism: Community, Identity, Continuity*, ed. Gowan Dawson and Bernard Lightman, 79–100. Chicago: University of Chicago Press.
Lloyd, J.T. 1970. Background to the Joule-Mayer Controversy. *Notes and Records of the Royal Society of London* 25 (2): 211–225.
Morrell, Jack, and Arnold Thackray. 1981. *Gentlemen of Science: Early Years of the British Association for the Advancement of Science*. Oxford: Clarendon Press.
Morus, Iwan Rhys. 2011. What Happened to Scientific Sensation? *European Romantic Review* 22 (3): 389–403.
Paul, Herman. 2014. What Is a Scholarly Persona? Ten Theses on Virtues, Skills, and Desires. *History and Theory* 53: 348–371.
Roberts, Robert J., and W. Jay Wood. 2007. *Intellectual Virtues: An Essay in Regulative Epistemology*. Oxford: Oxford University Press.
Smith, Crosbie. 1998. *The Science of Energy: A Cultural History of Energy Physics in Victorian Britain*. Chicago: University of Chicago Press.

Stanley, Matthew. 2008. The Pointsman: Maxwell's Demon, Victorian Free Will, and the Boundaries of Science. *Journal of the History of Ideas* 69: 467–491.

———. 2015. *Huxley's Church and Maxwell's Demon: From Theistic Science to Naturalistic Science*. Chicago: University of Chicago Press.

———. 2017. *Religious and Scientific Virtues: Maxwell, Eddington, and Persistence*. In this volume.

Tollebeek, Jo. 2011. *Men of Character: The Emergence of the Modern Humanities*. Wassenaar: Netherlands Institute for Advanced Studies in the Humanities and Social Sciences.

Yamalidou, Maria. 1999a. John Tyndall, the Rhetorician of Molecularity, Part One: Crossing the Boundary towards the Invisible. *Notes and Records of the Royal Society of London* 53 (2): 231–242.

———. 1999b. John Tyndall, the Rhetorician of Molecularity, Part Two: Questions Put to Nature. *Notes and Records of the Royal Society of London* 53 (3): 319–331.

Zagzebski, Linda Trinkaus. 1996. *Virtues of the Mind: An Inquiry into the Nature of Virtue and the Ethical Foundations of Knowledge*. Cambridge: Cambridge University Press.

# The Adventurer and the Documentalist: Science and Virtue in Interwar Nature Protection

**Raf De Bont**

**Abstract** The early twentieth century witnessed a growing sensibility about the decline of wild animal populations across the globe. In the interwar years, several international societies for nature protection were founded with the explicit goal to counter this trend by resorting to science. The societies in question set themselves the task to monitor the status of threatened animals through both fieldwork in faraway jungles and documentation efforts in the Metropolis. In this chapter, I argue that these two activities came with different scientific virtues. Preservationists associated fieldwork with ideals of endurance, truthfulness and discretion. They believed these ideals to be of utmost importance for acquiring knowledge in the wild habitats of endangered animals – which were presented as places of violent bewilderment and Otherness. In order to develop nature protection into a successful endeavor, however, fieldwork was not enough. Data coming from the field needed to be brought together, synthesized and made accessible. These activities – taking place in offices in cities of Western Europe and North America – called for other virtues. When these activities were discussed among preservationists, it was mostly stressed how they particularly required patience and precision. Thus, language of virtue was closely tied up with issues of place. The virtues of performing science in the field, after all, were not those of processing data in the documentation centre. It is by exemplifying the virtues of the field that (male) naturalist-explorers took on a public role as heroes of science. The patient and careful work in the documentation centre, to the contrary, was largely conceived as universalizing, disembodied and domestic. The people who carried out this work (mostly women) did so largely out of the public eye.

**Keywords** Epistemic virtues • Aims of science • Scientific practices • Nature protection • Fieldwork • Scientific hunting • Carl Ethan Akeley • Jean-Marie Derscheid • Tordis Graim

---

R. De Bont (✉)
Maastricht University, Maastricht, The Netherlands
e-mail: r.debont@maastrichtuniversity.nl

 
J. van Dongen, H. Paul (eds.), *Epistemic Virtues in the Sciences and the Humanities*, Boston Studies in the Philosophy and History of Science 321,
https://doi.org/10.1007/978-3-319-48893-6_9

## Introduction

> *Genius is innate. It is marked by the yet little understood gifts of superior insight, imagination, inventiveness, and power to carry to conclusion. It seems to bear no relation to cultural background, or subsequent milieu, although it is unlikely that genius ever flowers without the painful, even agonizing, dedication and toil with which lesser men must likewise pave their way to achievement* (Murphy 1964: 307–308).

The above citation is taken from an article that commemorates the centennial of the birth of Carl Ethan Akeley (1864–1924) – a reputed American taxidermist, naturalist, curator and preservationist, who worked for the American Museum of Natural History. The quotation mixes two well-known stereotypes from the genre of scientific biography. It presents Akeley as a truth-seeker with an exceptional set of inborn talents while at the same time stressing that he could not have reached his goals without blood, sweat and tears. The latter stereotype, highlighting the scientist's dedication and endurance, constitutes a recurring motif in commemorative writing about Akeley. It seemed fitting for a man who died in the jungle of the Belgian Congo while on expedition for the museum.

In recent historiography of science, quotations such as the one above are often analysed with reference to 'epistemic virtues' (Daston and Galison 2007). The concept of epistemic virtue has indeed proven useful in unravelling the ways in which norms influence shifting ideals of knowledge-making. Yet, scholars have also pointed out some caveats. In a recent article Creyghton et al. (2016) highlight the problem of demarcating exactly what is 'epistemic' in the virtues scholars and scientists cultivate. Of course, scientists hope to reach objective truths, but they also pursue other goals, including, for instance, those of a political or social kind. To attain the latter they invoke virtues that do not seem to differ much from those that virtue literature deems 'epistemic' (such as 'industriousness', 'meticulousness' and 'intellectual courage'). Thus, Creyghton *et al.* conclude, epistemic virtues are far from clearly delineated categories, embedded as they are in a wider and multiform normative landscape. In this chapter it is this embeddedness that I want to explore further.

By focusing on the blurred boundaries of epistemic virtue, I hope to broaden the discussion along three lines of inquiry. The first stems from the observation that the goal of science is not by definition limited to finding the truth. Of course, scientists hope to make advances in the immaterial sphere of knowledge, but often they *also* want to achieve something in the material world in which they live. As Robert Proctor (1991) has indicated, modern science's commitment to truth-finding does not necessarily entail value-neutrality. Science often pursues societal goals like world peace, economic stability or keeping the rise in global temperature under two degrees Celsius. Some sciences – such as counter-insurgency theory or conservation biology – are entirely organized around such goals. In other disciplines the societal aims are less tangible, but they never seem to be entirely absent. Thus, the virtues which scientists cherish and internalize also probably tie in with their ambitions to shape the world in particular ways.

Secondly, I believe we can broaden the *types of practices* we research when exploring the virtue language that scientists cultivate. Historians interested in epistemic virtues have particularly focused on those practices considered to be at the core of modern science: the acts of observation, experiment, or – in the humanities – textual criticism (Daston and Galison 2007; Paul 2011; Savoia 2015). As historians of various stripes have made clear, however, science also involves many other kinds of activities. The work of nineteenth-century biologists, for example, was self-evidently not limited to observation or experiment. As a genuine part of their science, biologists travelled across oceans; they gave lectures, shot animals and wrote letters; they collected, prepared and packed specimens; and they stored, classified and managed dead organisms, cell tissues or protein sequences (e.g. Strasser 2012). Such activities were (and are) vital to science, and their performance presumably went hand in hand with particular internalized virtues – epistemic or otherwise.

Lastly, the specific type of virtue language that emerged in the nineteenth and early twentieth-centuries also requires broader investigation. In Daston and Galison's narrative this is the period in which mechanical objectivity was held in highest regard and in which men of science hoped 'to eliminate themselves from observation' (2007, p. 96). By embodying virtues of self-effacement, so the argument goes, men of science hoped to create *dis*embodied knowledge (see also Levine 2002). Such striving for universal and 'placeless' truths was, of course, exemplified in the most iconic scientific site of the period: the laboratory. Yet, other places of scientific activity were to be associated with *different* ideals. Many scientists working in the field, for instance, stressed the importance of locality over universalism and the physicality of individual experience over disembodied abstraction (Livingstone 2003; De Bont 2015a). Such predilections surely also translated in specific scientific personae and particular forms of virtue language. Not every site of knowledge production, after all, calls for the virtues of the laboratory-worker.

In order to explore these three lines of inquiry, this chapter will focus on one case: the 'scientific' nature protection of the early twentieth century. Just as in present-day conservation biology, work in early-twentieth-century preservation science pursued a societal goal: research and the knowledge it generated was to serve the efficient protection of nature. Yet, while preservation legitimized science, the reverse was also true. The production of scientific knowledge was seen as one of the key goals of nature preservation. In this line of reasoning, wild nature was to be protected because it offered a crucial site of knowledge production. In the legitimation of both preservation and its science, the epistemic and the moral were connected in multiple ways and implicated in a wide range of activities – activities that went beyond the practices of observation and experiment and that took place in different spatial settings. Two in particular will be singled out in this chapter. I will start by exploring the virtues associated with the fieldwork carried out in the far-off 'wilderness'. As a point of contrast, I will then look – more concisely – at the largely hidden work of processing and synthesizing field data in the metropolitan office. As will become evident, the office and the wilderness produced very different scientific selves and virtues.

## A Science of Nature Protection

The focus of this chapter is on a science that, in the early twentieth century, was still in the making. Unlike contemporary conservation biology, nature protection science of this period did not constitute an institutionalized discipline. There were no research journals devoted to it, nor were there university chairs or departments. Even the idea itself that nature protection was to be a scientific endeavour did not go uncontested. In preservationist circles around 1900, it was aesthetic, moral and economic categories, rather than scientific ones, that were most often referenced to legitimize preservation work. Only in the 1910s and 1920s did naturalists start to put their mark on nature protection discourse. It is in the same years that preservation – thus far largely organized along national and local lines – became progressively more internationalized. Societies, committees and associations were set up that combined global with scientific ambitions. International conferences on nature protection were held at which scientists set the tone (in Paris 1923 and 1932), and conventions were negotiated, such as the London Convention on the protection of African fauna (1933) for which the input of scientific experts was deemed crucial. It was thus not at universities or research institutes, but at society meetings, conferences and convention negotiations that the science of nature protection took shape (Barrow 2009, 135–200; De Bont 2015b).

The agenda of this scientific nature protection was pushed by a small informal network of people largely to be found in a few cities in Western Europe and on the American East-Coast. This network partially consisted of professional naturalists employed by natural history museums, zoos or national parks. Yet, a substantial part of the network also comprised wealthy amateurs with an interest in collecting, hunting, animal breeding and bird-watching (De Bont, Schleper and Schouwenburg, forthcoming). Within this heterogeneous group, then, the idea took shape that scientific knowledge was needed to give nature protection a rational basis. And such scientific knowledge was mostly believed to be of an inventorying kind. Early-twentieth-century preservationists wanted to record the numbers and spread of threatened animals, their feeding and breeding habits and their migratory routes. They further aspired to locate and monitor the remaining 'wild' places on earth and map their soils and vegetation. They aimed to assess the impact of various forms of hunting and wildlife trade as well as agriculture and forestry. And because they hoped to rationalize international legislation and the management of nature reserves, they also started to compile inventories of existing legal systems and protected areas. All this meant that the science of nature protection largely revolved around paper technologies: lists, inventories, index cards and maps (Anonymous 1930; De Bont 2015b).

The creation of these paper technologies relied on both expeditions to the far-off wilderness and steady bookkeeping work in documentation centres. It was the first, however, that attracted by far the most public attention and it was also the naturalist-explorers engaged in these expeditions who provided early international nature preservation with its most iconic leaders and moral exemplars. The aforementioned

Carl Akeley offers a case in point. His untimely death meant that his involvement in the burgeoning movement of international preservation remained relatively limited, but his example proved inspirational for the generation that followed him. In one of his obituaries he was presented as the 'beau ideal of the naturalist explorer' (Roosevelt 1927). Over the following decades this ideal would be explicitly cultivated in preservation circles.

Several leading figures of the early twentieth-century movement followed in Akeley's footsteps, making their reputations as explorers. A good example is offered by the Belgian zoologist Jean-Marie Derscheid, who joined Akeley on his last Congo expedition in 1926, and who took over its leadership following the latter's death. In the years after the expedition, Derscheid became one of the leading preservationists on the European scene, setting up an International Office for the Protection of Nature and acting as its first director (Brien 1971). In the United States a similar role was taken up by the primatologist Harold Jefferson Coolidge, who became renowned both for his expeditions to Central-Africa and Indo-China and for his continued efforts to stimulate American involvement in international wildlife protection (Cevasco 1997). Lastly, international bird protection was also closely linked with naturalist exploration. Its most active European promotor was Jean Delacour, a French gentleman-scientist and bird collector who carried out numerous expeditions to South-East Asia and Africa (Mayr 1986). As a group these men serve as a good entry point to study the virtue culture of scientific nature protection in the interwar years.

Derscheid, Coolidge and Delacour worked in close consultation with each other and pursued similar interests. Like Akeley, all of them spent time in the field studying threatened species, with a particular interest in their number, geographical spread, behaviour and morphological variety (see e.g. Derscheid 1927a; Coolidge 1930; Delacour and Jabouille 1931;). All of them deemed these issues important for science itself, but also for furthering nature protection. And all of them took their expedition work as central to their personal identity. In particular, their self-presentation and aura relied heavily on the trope of belonging to the *last* generation of explorers. Men such as Akeley were staged as adventurers who filled the last blank spaces on the map and, thus, were the last to encounter truly unspoiled wilderness (Roosevelt 1927; Derscheid 1928, 94–95; Coolidge and Roosevelt 1933, 13–15). Looking back on his expeditions in his autobiography, Delacour expressed this idea with a combined sense of pride and nostalgia: 'It was my good luck to explore the last large region that had remained almost unknown in my particular field. Such possibilities do not exist anymore' (Delacour 1966, 83).

The 'discovery' of the last places of unspoiled wilderness came with particular moral obligations. After all, the fear was widespread that the act of discovery itself would open up the last patches of wilderness to civilization and in this way threaten their very existence.[1] The description of newly discovered wilderness was therefore often complemented by the immediate call for its protection (Derscheid 1927b; Delacour 1931). In this context the relationship of the explorers themselves to the far-

[1] For more on this trope, see Woebse, 2011, 45.

off and wild locations they were penetrating was thematized explicitly. Engaging with such places (which were presented as both threatening and threatened) was believed to require a balanced series of skills and virtues. It is to those that I will now turn.

## The Field (1): Endurance

Understanding the far-off wilderness asked for men who were willing to take risks and endure physical discomfort. In the introduction to Akeley's best-selling *In Brightest Africa* (1923), the director of the American Museum of Natural History, Henry Fairfield Osborn, stressed exactly that. Akeley, Osborn insisted, took the risk-taking so far that he had been 'close to death at three occasions'. Osborn referred, amongst other things, to attacks by a leopard, a lioness and an elephant – attacks Akeley himself recounted in detail in the book *In Brightest Africa* (Akeley 1923, X–XII). Stories of the assaults, accompanied with photographs of a wounded expedition leader, were a recurring image in the self-presentation of Akeley (Akeley 1923, 30–31, 39–57, 97–102), and they were also to reappear endlessly in the biographical writings that his wife Mary J. Akeley devoted to him after his death (e.g. Akeley 1944, iii, 56–59, 117–122). The images of an injured Akeley were more than a simple proof of physical heroism; they performed an epistemic function as well. Much like the frost wounds of polar explorers that were similarly the object of media attention in the same period, injuries caused by animal attacks served as 'visceral testimony to having "been there"' (Herzig 2005, 78). They were markers of scientific credibility. Akeley stressed that, in order to understand wild animals as they really were, it was necessary to get in 'closest touch' with them (Akeley 1923, 263). His expeditions, he made clear, were not pleasure hunts but scientific endeavours: 'Had I been hunting merely to kill, I should have laid off a day. But science is a jealous mistress and takes little account of a man's feelings' (Akeley 1923, 211).

Yet, Akeley's stories about the confrontation with dangerous animals were ambiguous. While relying on old tropes, Akeley was also bent on transforming them. Starting with the elusive mountain gorilla, the taxidermist began to represent wild animals as threatened rather than threatening. It is in this transformation that the preservationist logic takes off. Against earlier images of ferocity, Akeley framed the mountain gorilla as a placid creature – a point which he stressed by pointing out that his expedition was joined by 'ladies without previous hunting experience' (Akeley 1923, 196). The highly gendered narrative strategies of Akeley and his biographers have already been elaborately analysed by Donna Haraway (1989, 46–52). In the context of this article, it can be added that they conform nicely to the ambiguities of the virtue language surrounding field expeditions. Women were written out of the story when Akeley's masculine endurance was of epistemic importance; they were written back in to it when preservationist norms required so.

Most international preservationists followed Akeley in casting doubt on the ferocity of wild animals. Looking back on his expeditions of the 1920s and 1930s, Delacour indicated that 'dangers due to so-called 'savage' animals […] [were] almost negligible' (1966, 84). Yet, in itself such statements did little to alter the persona of the expedition leader. After all, plenty of other inconveniences remained to uphold the trope of endurance. In his autobiography Delacour listed: 'malaria, typhus and dysentery [...] oppressive heat [...] mosquitoes and other insects [..] tiresome transport [...] [and] bother with staff, often belonging to various human types and ready to quarrel' (1966, 85). A similar list can be found in the field notes which Derscheid published as an odd mixture of scientific report and autobiographical writing: 'hunger, thirst, frost, storm, tempest, rain, mutiny among the porters, bleeding feet, nails torn on the rocks. And on top of that all: the tragic death of the expedition leader, caused by an infernal climate' (Derscheid 1927b, 35).[2]

Once home, explorers continued to refer to tropical illnesses like malaria and other marks of physical discomfort. They counted as signs of past endurance. References to such discomfort appeared in field reports, autobiographies, obituaries, and interviews – but also in more private sources such as personal correspondence.[3] In an emotional letter to Coolidge, for instance, Richard Strong brought up memories of their joint expedition to Africa. Coolidge, Strong recalled, had shown the most courage of the expedition's staff, as demonstrated by the fact that during the expedition he had become so 'very thin'. Field exploration, Strong believed, constituted a 'severe test of human nature' and Coolidge had passed it. Even in the most distressing of circumstances, after all, he had remained his kind self. References to physical suffering, this example illustrates, not only played a role in creating scientific credibility, but also in male bonding between field scientists. Endurance was not only an epistemic, but also a social virtue.[4]

Among preservationist-explorers virtues and skills were usually evaluated in their physical rather than their intellectual guise. When a certain Julian Felz contacted Coolidge in 1932 with the question of whether he could be hired for an upcoming expedition, he wrote the following:

> I am thoroughly familiar with travel in the tropics, know how to doctor malaria, dysentery etc., and as I was nearly three months in the sleeping sickness belt of the Luftwanyana, and came out alive, I seem more or less salted. ... As far as shooting is concerned, I have steadily qualified expert on the US army qualifications. I know rifles and pistols and have hunted on three continents.[5]

Felz added that he had 'also done some scientific publishing', but that he was not going to bore Coolidge with a list. It was, after all, not the disembodied criteria of science on paper that mattered, but the embodied abilities acquired in specific places.

---

[2] Similar examples of discomfort can also be found in Coolidge's expedition report from Indo-China (Coolidge and Roosevelt 1933).

[3] See, for instance: the correspondence of Coolidge and Derscheid with Pieter-Gerbrand van Tienhoven, Archief van de Nederlandse Commissie voor Internationale Natuurbescherming, Amsterdam City Archives, (ANCIN), 1283–50 and 55; transcript of interview with Harold Coolidge, 27 June 1978, as part of the Boone and Crockett Club, oral history project, HUG (FP) 78.10, Box 40, Harold Jefferson Coolidge Papers, Harvard University Archives, Cambridge, MA (HJCP).

[4] Strong to Coolidge, March 23, 1944, HUG (FP) 78.10, Box 17, HJCP.

[5] Felz to Coolidge, April 2, 1932, HUG (FP) 78.10, Box 17, HJCP.

Felz's application is interesting not only for its contents but also for its tone. Felz used the understated bravado of the practical man that one also encounters in the published work of the preservationist-explorers discussed in this chapter. Typically, these explorers used wry humour to depict their confrontation with inhospitable yet enchanting nature (e.g. Akeley 1923, 49–51; Coolidge 1928; Derscheid 1927a, b).[6] In this way their persona was constructed in parallel to the natural spaces that needed protection. After all, virtues of willpower, self-reliance and endurance became visible in the confrontation with hostile and primitive jungles as well as in ambiguous encounters with animals of strong corporeal charisma such as the gorilla or the leopard (cf. Lorimer 2015). The descriptions of these encounters were not only highly gendered, but also strongly racialized. It is active white men that were believed to embody the virtues necessary for confronting wild nature, while Africans were at best represented as 'useful technical staff' (Delacour 1966, 90) or at worst as 'boys' who run off in the face of danger (Coolidge 1928, 98) and 'weep like children' when confronted with harsh conditions (Derscheid 1927a, b, 37). White expedition leaders, however, were believed to maintain their good humour under all circumstances. After enumerating all the physical annoyances of field expeditions in the Congo, Derscheid, for instance, wrote in his diary:

> At the end of the day, I enjoy this life as I enjoy the flesh of a freshly shot antelope! I drink from the same cup as a negro, who, within an hour, I will beat with the sjambok – illegally of course – and I will chase gorillas with no arms other than a small revolver and a camera (Derscheid 1927a, b, 36).

The quote is written in a tone of self-conscious brutality. In this way the field is constructed as a place with rules different to those of the civilized world. Later in his published diary Derscheid explicitly stated that 'everything here is unexpected, violent, different' (Derscheid 1927a, b, 47). It was in dealing with places of violent bewilderment and Otherness that the preservationist-explorer could showcase his virtues. Understanding the local specificities of the far-off wilderness demanded a particular persona – brave, brutal and masculine.

## The Field (2): Truthfulness and Discretion

Working in the field – with its risks, its physical challenges, but also with its moments of enchantment – was systematically presented as an adventurous undertaking. The term 'adventure' recurs in the written work of Akeley, Coolidge, Delacour and Derscheid and as an idea it played an important role in their self-presentation and ethos. Yet, while the association with adventure was actively sought (and easily translated into public attention), it also came with qualifications. The preservationist-explorers discussed in this chapter sought a *scientific* aura and

[6] In a letter to Van Tienhoven, Derscheid described the Congolese region he was exploring as 'the most marvellous of sites' in 'the most depressing of climates', March 11, 1927, ANCIN, 1283–55.

did not want to be conflated with the 'sensationalists' who published travelogues or hunting stories, set up spectacular exhibitions or shot Hollywood movies. Given the fact that they addressed, at least in part, the same audiences, this implied a subtle kind of boundary work.

In their attempts to distinguish themselves from non-scientists, preservationist-explorers put a lot of emphasis on the virtue of 'truthfulness'. In his introduction to Akeley's *In Brightest Africa*, Osborn explicitly wrote that it was truthfulness that comprised 'the high note of the biographer of animal life'. Akeley, he added, could be considered the embodiment of such truthfulness. In his taxidermist reconstructions of African animals Akeley was said to 'show them as they really are, not as they are shown by sensationalist writers' (Akeley 1923, xi). Akeley himself insisted that his taxidermy was 'scientifically accurate' because he had observed the animals he reconstructed in the wild. He stressed this made him quite different from taxidermists who worked from specimens brought to them by hunters (Akeley 1923, 264). Again, it was first-hand and emplaced experience – as testified by Akeley's bodily suffering – that made him worthy of trust.

Yet, 'having been there' was not sufficient. After all, sensationalist stories were often produced by men who clearly *did* have local field experience, but who were believed to have their reasons to distort the truth. Derscheid for instance denounced the publicist and filmmaker Benjamin Burbridge as a '*coureur d'aventures*', who exaggerated his estimation of the number of mountain gorillas in order to get the permission to shoot and trade more of them (Derscheid 1927a, b, 71). Coolidge was likewise critical of the estimates made by Martin Johnson, director of the successful movie *Congorilla.* In his letters Coolidge criticized Johnson as someone who believed himself to have discovered gorillas in a particular region of the Belgian Congo and who therefore exaggerated their number as a manoeuvre of self-aggrandizement.[7] It was not the fact that men such as Burbridge and Johnson were involved in wildlife trade and movie productions that was seen as problematic. After all, Derscheid and Coolidge themselves traded in exotic specimens and they also used films for their public talks. What they criticized was the overtly commercial aspect of enterprises like Burbridge's and Johnson's. In a letter to Coolidge, Derscheid extensively complained that both had used science only as a pretext for their expeditions, but that their real goal had been financial gain.[8] In short, embodied experience of specific localities and their wildlife did not suffice. One needed to have the discipline to convey truthfully such experience to audiences at home. Commercial intentions, so it was suggested, undermined this discipline.

The naturalist-explorers of the early preservation movement made clear that they did not to belong to the commercial class. They were linked to scientific institutions, and – besides that – several had personal fortunes that kept them even 'independent' from these. Derscheid sprung from a wealthy family and was not paid for his preservation work. Delacour inherited a family fortune and lived as a pseudo-aristocrat with his own chateau and a private menagerie. And while Coolidge proved to inherit

[7] Coolidge to Mary Jobe Akeley, February 1, 1932, HUG (FP) 78.10, Box 17, HJCP.

[8] Derscheid to Coolidge, July 1, 1932, ANCIN, 1283–50.

less than he expected when his father died, he still partially undertook financial responsibility for his own expeditions.[9] Independence of means was believed to come with independence of mind and thus counted as a facilitator of epistemic virtue. To quote Delacour: 'the urgent necessity of making money, which has poisoned the lives of so many promising young men, did not exist in my case' (Delacour 1966, 22). Just like having suffered through a malaria infection, inherited wealth was a sign of credibility for naturalist explorers.

The way truthfulness was embodied among preservationist-explorers in the 1920s and 1930s illustrates the ways in which epistemic virtues are socially embedded. Drawing at least in part on old ideals of the gentleman-scientist, being truthful was not only framed as a matter of personal character, but also as a matter of social class.[10] Truthful behaviour was scientifically rigorous *as well as* gentleman-like. Hence, while an adherence to the truth might seem the most quintessentially 'epistemic' of virtues, it was strongly shaped by norms that were imported from non-scientific contexts.

The same argument can be made about the internalized norms that regulated the preservationist-explorer's interaction with wild animals. Akeley, Derscheid, Coolidge and Delacour all *hunted* as part of their scientific work. In their ideas about what comprised 'good' hunting practices they were strongly influenced by the codes of so-called 'sportsmanship' that had developed in circles of British aristocrats and the industrial American elite around 1900. Historians have discussed these codes in detail, exemplified, for instance, by former US president Theodore Roosevelt and propagated in the elitist Boone and Crockett Club he founded (Dunlap 1988; MacKenzie 1988; Jones 2016). Akeley, Derscheid, Coolidge and Delacour were all closely associated with Roosevelt's networks, and were influenced by the same moral of increased self-regulation when it came to hunting. With the members of the Boone and Crockett Club they shared an abhorrence of the commercial hunt, and like them they stressed the importance of endurance, moderation and humanity in confrontation with individual animals – which were presented as equals that should receive a fair chance (e.g. Akeley 1923, 155; Coolidge 1928; Derscheid 1927a, b, 42–43).

While they drew on codes of behaviour that developed in elite hunting circles, Akeley and his fellow preservationist-explorers did not identify as 'sportsmen' nor were they perceived as such. In their rationale of the hunt, science came first. In his obituary of Akeley, the Belgian ambassador to the US, Émile de Cartier de Marchienne, wrote for instance: 'Although [Akeley] [...] was counted "a mighty hunter", he never killed for the sake of killing. He could kill wild beasts for protection, for food, or for the legitimate purposes of science, but his soul revolted against

[9]Coolidge to Van Tienhoven, November 8, 1938, ANCIN, 1283–50; John C. Phillips to Van Tienhoven, August 19, 1934, ANCIN, 1283–110.

[10]This arguably played out somewhat differently in the US than it did in Europe, where there was a long-standing tradition of gentleman science dating back to early modern times. Yet, criticism of the dangers of commercial science were definitely not absent in the US. See e.g. Shapin 1994; Lucier 2009.

the wanton destruction of innocent animals or rare species whose conservation is necessary for scientific study' (Cartier de Marchienne 1927). The double use of science is of importance here. While science served as a legitimation for hunting, its long-term perspective also constituted an important limitation. Hunting provided the material for museum exhibits as well as for taxonomic and morphological research, but it could also threaten the very object of both nature preservation and its science. The scientific hunt was therefore a balancing act, only allowed when it advanced knowledge and in no way hampered it in the long run. A virtuous preservationist-explorer *did* hunt, but moderately and reluctantly.[11]

Whether somebody abided by the codes of scientific hunting was often a matter of interpretation. Despite his preservationist aura, Coolidge, for instance, was to become embroiled in a controversy over his hunting behaviour. Interested in the study of variation within species, he had set up an expedition in 1937 which 'collected' more than 100 gibbons in South East Asia. This high number led to criticism within the preservation movement, but Coolidge defended his credentials as a nature protector, stating that his collecting did not threaten the population and was of great scientific importance. To his critics in Europe he expressed the 'definite feeling that a person who has hunted or collected in the field can, in many ways, evaluate certain wild life problems better than one who has not the advantage of field experience.'[12] The American Committee for International Wildlife Protection, of which Coolidge was one of the co-founders, had always stressed that 'specimens of rare and disappearing species should be collected with discretion'.[13] Such discretion, Coolidge was now arguing, should be rooted in first-hand and place-based knowledge. In this case virtue also came with a well-developed sense of place.

It was clear that, despite Coolidge's claim of discretion, hunting incidents were potentially harmful for a preservationist's reputation. When, in the postwar years, Coolidge set up new expeditions under the umbrella of his Coolidge Foundation, he accordingly took precautions and made all participants sign a document that read:

> He [the participant] … agrees to recognize that the present field operation is strictly dedicated to scientific research and to the high moral principles of service to mankind and the increase and discussion of knowledge among men. … The Coolidge Foundation, being devoted to the above principles, wishes to protect itself and its employees and associates from the possible criticism that it is using its good name to foster a hunting party.[14]

If not internalized, the ethos of discretion was to be enforced.

---

[11] As such they were often involved in lobbying for more severe hunting laws in areas where they actively hunted themselves. In a letter to his American friend William Hornaday, Delacour, for example, bragged about the skins he collected in South-East Asia (of more than 300 species!), while including a copy of his request to French authorities there demanding stricter regulation. The time had come, he told Hornaday, 'to stop the silly shooting'. Delacour to Hornaday, July 7, 1924, William Hornaday Papers, Coll. Nr. 1001, Wildlife Conservation Society Archives, New York.

[12] Coolidge to Van Tienhoven, February 25, 1939, ANCIN, 1283–50.

[13] American Committee for International Wildlife Protection, 12 Dec. 1935, Secretary's report, ANCIN, 1283–110.

[14] Document sent with a letter by Coolidge to Roger Heim, May 26, 1952, Roger Heim Papers, Box 46, Muséum d'Histoire Naturelle, Paris.

In epistemological literature discretion is often mentioned as an epistemic virtue (e.g. Zagzebski 1996). With regard to scientific hunting, however, it is not only a norm that regulates truth-finding, but also one that accords with the ideology of nature protection. Acquiring knowledge – about, for instance, gibbon variation – was crucial, but also seen as secondary to the ultimate goal: preserving threatened nature.

Akeley, Coolidge, Delacour and Derscheid continuously used virtue language when writing about themselves – and others did so when writing about them. In their personae, as I have shown, virtues such as endurance, truthfulness and discretion were central categories. These, however, were not just virtues of the placeless mind, exclusively directed towards regulating intellectual conduct. They were virtues that could only be acquired (and proven) in bodily interaction with particular places. These virtues, furthermore, perpetuated norms from the social milieus from which the preservationist-explorers sprang and converged with the societal ends of the men's science.

## The Office: Patience and Precision

Science-based preservation required more than individual field reports from specific places. It also demanded that the data be synthesized and made accessible to be used to further the cause of nature protection worldwide. In the interwar years, preservationists set up international organizations to do exactly that. Coolidge, as indicated, was one of the founders of the American Committee for International Wildlife Protection (Barrow 2009, 135–200). Delacour, for his part, was engaged in organizing the International Committee for the Protection of Birds (Pearson 1937, 370–419). And Derscheid, finally, was the first director of the International Office for the Protection of Nature, which was established in Brussels in 1928 (Pelzers 1994). It is to this last organization that I will now turn in order to explore whether the metropolis generated different scientific virtues from those of the wilderness.

The self-set goal of the Brussels Office was to collect data useful for the global protection of nature. These included information on threatened animals, but also on existing hunting laws and nature reserves throughout the world, or on the whereabouts of travelling hunters who showed unsportsmanlike behaviour. The office was founded on the basis of the twin ideals of universalism and synthesis. It was to collect and make accessible global information – as it was hailed in the utopian language of the rising documentation science of the time (Rayward 2008). The link with such documentation science gave the Office an aura of modernity, but in essence it used quite traditional tools to acquire knowledge: a library, index cards and correspondence. The Office collected information through scientific and legal periodicals, from newspaper reports, by sending out questionnaires, and by corresponding with informants in the field. The staff answered preservation-related questions when contacted, and helped to prepare expeditions (Anonymous 1930; Pelzers 1994). As such it was definitely a place of knowledge creation and dissemination, but it constructed this knowledge around a different ethos than that prevalent in the field. This becomes very clear when we look into the criticism Derscheid encountered concerning his work at the Office.

Relatively early on, some heavyweights in the nature protection movement expressed doubts as to whether Derscheid was the right man for the job. While his expeditions were consistently praised, he was also described as 'high strung', 'nervous' and 'lacking tact'. This might not have been such a drawback in the Congolese jungle, but it was considered problematic in a documentation centre in Brussels.[15] Gradually, complaints would arise over Derscheid's lack of punctuality and accuracy in handling both the Office's correspondence and its administration.[16] This was seen not just as a want of courtesy on his part, but also as an obstacle to setting up a knowledge network and, thus, as a major hinderance to the actual functioning of the Office. After all, it hampered the very gathering and dissemination of information the Office was supposed to facilitate. Ultimately, persistent complaints led to Derscheid's removal from his post.[17] He would be replaced by his female assistant: the Norwegian Tordis Graim.

Trained in law, Graim worked as head of the documentation section of the Office, until, in 1934, she was asked to take over Derscheid's position. Coolidge described her in a letter as 'of more than usual competence' and many other preservation leaders concurred.[18] When recounting her strengths, they referred not only to her legal training and her knowledge of five languages, but also to virtues such as precision, patience, and conscientiousness.[19] Of course, these virtues differ substantially from those cultivated among naturalist-explorers. Graim also made sure at all times to present herself as a compiler and provider rather than a heroic *producer* of knowledge. Since she did not claim to possess first-hand experience of particular places or threatened animals, she did not have to show proof of her capability of physical endurance. Her type of work was performed at a safe distance from the natural spaces in which preservationists were interested. What was expected from her was a careful and swift processing of information provided by others.

These expectations strongly guided Graim in her behaviour both inside and outside the Office. When still working as head of the documentation center she refused to replace Derscheid as speaker at an international conference, as she lacked first-hand experience on the topic he was to address: African nature. She did want to prepare a paper on the functioning of the Office, but even then believed it better that somebody else deliver it.[20] While over the years she attended and prepared several

[15] Phillips to Van Tienhoven, August 14, 1928, ANCIN, 1283–110.

[16] Victor Van Straelen to Van Tienhoven, June 9 and Dec. 7, 1933, ANCIN, 1283–127; Van Tienhoven to Coolidge, June 17, 1933, ANCIN, 1283–50.

[17] See e.g. the extensive correspondence on this topic between Van Straelen and Van Tienhoven throughout the years 1933–1934, ANCIN, 1283–50.

[18] Coolidge to Van Tienhoven, November 25, 1932 ANCIN, 1283–50.

[19] These appreciative leaders (originally) included Derscheid. See e.g. Derscheid to Van Tienhoven, January 20, 1929, ANCIN, 1283–55; Van Tienhoven to Delacour, August 2, 1935, ANCIN, 1283–35.

[20] Graim to Van Tienhoven, June 8, 1931, ANCIN, 1283–64.

conferences and meetings, her letters also testify to a strong hesitation on her part to participate in the actual discussions. She was happy to bring together factual material to facilitate a conference, but never assumed the role of authoritative 'specialist'.[21] Her knowledge was always second-hand. Yet, she would also insist that such knowledge was more than sufficient for her work at the Office. This explains why she was greatly insulted when John C. Phillips, an influential American preservationist, suggested that, instead of Graim, the Office should hire a zoologist to compile red lists of threatened species. Such work, Graim stressed, did not necessitate direct zoological experience; it required a broad background as well as care and patience – virtues she had amply shown she possessed.[22]

Within the Office, patience, care and precision were cherished virtues not only for epistemic reasons. While such qualities enabled researchers to get a truthful sense of the state of threatened nature, they simultaneously served other ends. Notably, they were deemed crucial to binding informants as well as fundraisers and governments to the Office. Creating an efficient form of nature preservation ultimately demanded such contacts. Victor Van Straelen, one of the Office's council members, argued that documentation work had to be carried out in an 'orderly and businesslike' manner in order to fulfil the Office's preservation task efficiently.[23] In short, patience, care and precision were not just useful for generating truth; they were essential to the successful realization of the societal goals of nature protection.

Like the virtues of the naturalist, the norms cherished at the Office seemed to be highly gendered. It is in any case striking that in other institutional contexts functions similar to those that Graim performed were also carried out by women: by Phyllis Barclay-Smith at the International Committee for the Protection of Birds, and by Elisabeth Hone and Anne Tilney Sage at the American Committee for International Wildlife Protection. This observation accords with the conclusions of, amongst others, Naomi Oreskes, who has stressed that, around 1900, objectified scientific tasks that asked for the effacement of the individual were increasingly performed by women (Oreskes 1996). The same period, furthermore, witnessed a strong feminization of office work and an increasing association of such work with patience, tact and domesticity (Strom 1992). In both science and administration women were considered particularly apt for tasks carried out outside the public sphere, such as those performed at documentation centres. Unlike the mediatized exploits of the naturalist-explorer, the documentalist's patience only left its traces in personal correspondence and index cards. The difference between fieldwork and office work could not have been greater. In the wilderness first-hand knowledge of specific places could only be acquired through brutality and bodily endurance, while the synthesizing and global knowledge of the office demanded tact and care. Graim's sort of work was disembodied; its practitioner replaceable.

[21] Draft letter to Louis Pittet, undated [1931], Graim to Van Tienhoven, June 20, 1934, ANCIN, 1283–64.

[22] Graim to Van Tienhoven, February 4, 1937, ANCIN, 1283–65.

[23] Van Straelen to Van Tienhoven, December 8 1932, ANCIN, 1283–127.

All this, however, does not mean that roles, virtues and personae within the documentation centre were homogeneous or clearly bordered entities. The aforementioned women that processed data and organized the information flow for the preservation movement were more diverse than they might seem at first glance – both in terms of their backgrounds and the virtues for which they were credited. Anne Tilney, for instance, performed a function quite similar to Graim's, but Coolidge particularly commended her for her 'unusual field experience' and for 'having hunted in Africa, India, Indo-China, China and Alaska'.[24] Phyllis Barclay-Smith combined her work at the Committee with publishing ornithological books, while also playing an increasingly visible and vocal role in the various preservationist organizations in which she was engaged. This eventually earned her the sobriquet of 'dragon' (Matthews 1980). Graim, for one, believed Barclay-Smith's ambitions testified to a 'misguided feminism'.[25] She stuck to a more traditional female role herself, in which virtues were deemed impersonal and tasks were to be undertaken with modesty. 'I really love to do work', she wrote in one of her letters, 'but I love to do it in the shadows'.[26] Graim's name is, not surprisingly, still absent from biographical dictionaries.

## Conclusions

When studying the internalized norms that guide scientists in their activities, it is not easy to discern the epistemic from the non-epistemic, the scientific from the non-scientific. This should not come as much of a surprise. After all, many practices carried out by scientists are not readily recognizable as 'scientific' and have clear equivalents in other fields of human activity. In this chapter, I have attempted to further this point by looking at a field from the margins of the world of science: the little-institutionalized scientific nature protection of the early twentieth century. I have, furthermore, focused on activities that, while crucial to nature protection science, are generally not seen as the most fundamental in an epistemological sense: travelling, hunting, writing letters, assembling lists of threatened species.

The first conclusion to be drawn from analysing such activities with a low epistemic status is that they have been the object of a highly developed virtue language. In fact, in the field of scientific nature protection it is arguably those activities that received *most* attention in the normative valuing of both the work of oneself and of others. The actual *observation* of threatened animals and landscapes received relatively limited consideration in commemorative articles, autobiographies, diaries or letters. It was largely the demanding period that took place before such observation

[24] Secretary's report, American Committee for International Wildlife Protection, 12 Dec. 1935, ANCIN, 1283–50.

[25] Graim to Van Tienhoven, July 6, 1936, ANCIN, 1283–65.

[26] Draft letter to Louis Pittet, undated [1931] and Graim to Van Tienhoven, June 10, 1934, ANCIN, 1283–64.

could even occur (in which one had to travel through dense jungles or chase hidden gorillas) and the somewhat dreary period afterwards (in which information was to be processed and exchanged) that were spoken of in moral and moralizing terms.

For disciplining the aforementioned activities, virtues were borrowed from (or at least shared with) non-scientific actors such as the elite hunter, the gentleman and the office worker. This testifies to the broader social embedding of virtues held dear in nature protection science. Yet, the fact that the people discussed here clearly shared virtues and ideals of the self with non-scientists does not detract from the fact that they often also felt the need to dissociate themselves from the latter. Men such as Coolidge and Derscheid had to make clear they were not sensationalist writers or recreational hunters, but true naturalists. And Graim would react bitingly in her letters when she was mistaken for a secretary or a stenographer.[27] She wanted to be seen as a documentalist, not an office worker.

The virtues of nature protection science took on particular forms. Yet, in several instances these virtues were shaped less by epistemic ideals than by the science's ideological ends. For both the naturalist-explorers and documentalists discussed here it was ultimately the efficient protection of nature that was at stake. This goal came with behavioural norms. It meant one had to hunt with discretion, deal carefully with correspondence, and synthesize data with patience and precision. Such norms, furthermore, were particularly made explicit in polemical contexts. When identities were challenged or authority was threatened, virtue language flourished. Coolidge started to write about his own discretion after accusations of over-enthusiastic hunting. And Graim thematized her own patience and care, when it was suggested that she could be replaced by a zoological specialist with first-hand knowledge of wild nature.

A final conclusion put forward by the cases discussed here is that virtue language is closely tied up with issues of place. The virtues of performing science in the field, after all, are not those of processing data in the documentation centre. The difference maps at least partially onto dichotomies between the public and the private, the particular and the universal, the embodied and the disembodied, the masculine and the feminine. Fieldwork was associated with virtues such as endurance, truthfulness and discretion and was framed in terms of individual and embodied encounters with particular wild places and wild things. It is by exemplifying these virtues that (male) naturalist-explorers took on a public role as heroes of science. The patient and careful work in the documentation centre, to the contrary, was largely conceived as universalizing, disembodied and domestic. The women who carried out this work did so largely out of the public eye. Some virtues of scientific nature protection were meant for the limelight. Others were to be performed in the shadows.

[27] Graim to Van Tienhoven, June 10, 1934, ANCIN, 1283–65.

# Bibliography

## *Primary Sources (Unpublished)*

Harold J., Coolidge Papers, Harvard University Archives, Cambridge, MA. Papers relating to expeditions (HUG (FP) 78.10, Box 17).

Pieter-Gerbrand van Tienhoven Papers., Archief van de Nederlandse Commissie voor Internationale Natuurbescherming, Amsterdam City Archives. Correspondence with Harold J. Coolidge (1283–50), Jean-Marie Derscheid (1283–55), Tordis Graim, (1283–64-66), John C. Phillips (1283–110) and Victor van Straelen (1283–127).

Roger Heim Papers., Muséum d'Histoire Naturelle, Paris. Correspondence with Harold J. Coolidge (box 61).

William Hornaday Papers., Wildlife Conservation Society Archives, New York. Correspondence with Jean Delacour (Coll. Nr. 1001).

## *Primary Sources (Published)*

Akeley, Carl. 1923. *In Brightest Africa*. London: Heinemann.

Akeley, Mary Jobe. 1929. *Carl Akeley's Africa*. New York: Dodd, Mead & Company.

———. 1944. *The Wilderness Lives Again*. New York: Dodd, Mead & Company.

Anonymous. 1930. *L'Office Internationale pour la Protection de la Nature: Ses Origines, Son Programme, Son Organization*. Brussels: Office International pour la Protection de la Nature.

Cartier de Marchienne, Émile de. 1927. Akeley, the Conservationist. *Natural History* 37: 115–117.

Coolidge, Harold J. 1928. Wanted – One Giant Gorilla. *The Sportsman* 4: 41–42 and 98.

———. 1930. Notes on the Gorilla. In *The African Republic of Liberia and the Belgium Congo, Based on the Observations made an the Material Collected during the Harvard American Expedition, 1926–1927*, ed. Richard Strong, 623–625. Cambridge, MA: Harvard University Press.

Coolidge, Harold J., and Theodore Roosevelt. 1933. *Three Kingdoms of Indo-China*. New York: Thomas Y. Crowell Company.

Delacour, Jean, and Pierre Jabouille. 1931. *Les Oiseaux de l'Indochine Française*. Vol. 4 vols. Paris: Exposition Internationale Paris.

———. 1966. *The Living Air: The Memoirs of an Ornithologist*. London: Country Life Limited.

Delacour, Jean and Pierre Jabouille. 1931. *Les Oiseaux de l'Indochine Francaise*. Paris: Exposition Internationale Paris. 4 vols.

Derscheid, Jean-Marie. 1927. Notes sur les gorilles des volcans du Kivu. *Annales de la Société Royale Zoologique de Belgique* 58: 149–159.

———. 1927b. *La Protection Scientifique de la Nature*. Brussels: Henri Kumps.

———. 1928. Entre le lac Édouard et le Kivu—Mission Akeley-Derscheid (1926–1927). *Bulletin de la Société royal de Géographie d'Anvers* 48: 92–99.

Herzig, Rebecca. 2005. *Suffering for Science in Modern America*. New Brunswick: Rutgers University Press.

Murphy, Robert Cushman. 1964. Carl Ethan Akeley. *Curator: The Museum Journal* 7: 307–320.

Pearson, Thomas Gilbert. 1937. *Adventures in Bird Protection: An Autobiography*. New York: D. Appleton Century.

Roosevelt, Kermit. 1927. Akeley, the Explorer. *Natural History* 37: 118.

## Secondary Sources


Barrow, Mark. 2009. *Nature's Ghosts: Confronting Extinction: From the Age of Jefferson to the Age of Ecology*. Chicago/London: University of Chicago Press.

Brien, Paul. 1971. Jean-Marie Derscheid. *Biographie nationale* 37: 211–235.

Cevasco, George A. 1997. Coolidge, Jefferson Harold. *Biographical Dictionary of American and Canadian Naturalists and Environmentalists,* ed. Keir Brooks Sterling et al., 166–168. Westport/London: Greenwood Press.

Creyghton, Camille, Pieter Huistra, Sarah Keymeulen, and Herman Paul. 2016. Virtue Language in Historical Scholarship: The Cases of Georg Waitz, Gabriel Monod, and Henri Pirenne. *History of European Ideas* 42: 924–936.

Daston, Lorraine, and Peter Galison. 2007. *Objectivity*. New York: Zone Books.

de Bont, Raf. 2015a. *Stations in the Field: A History of Place-Based Animal Research, 1870–1930.* Chicago/London: University of Chicago Press.

———. 2015b. Borderless Nature: Experts and the Internationalization of Nature Protection. In: Scientists' Expertise as Performance: Between State and Society, 1860–1960, ed. Evert Peeters, Joris Vandendriesche and Kaat Wils, 49–65. London: Pickering and Chatto.

de Bont, Raf, Simone Schleper and Hans Schouwenburg. Forthcoming. Conservation Conferences and Expert Networks in the short 20th Century. *Environment and History.*

Dunlap, Thomas R. 1988. Sport Hunting and Conservation, 1880–1920. *Environmental History Review* 12: 51–60.

Haraway, Donna. 1989. *Primate Visions: Gender, Race, and Nature in the World of Modern Science*. London and New York: Routledge.

Jones, Karen. 2016. *Epiphany in the Wilderness: Hunting, Nature and Performance in the Nineteenth-Century American West*. Boulder: University of Colorado Press.

Levine, George. 2002. *Dying to Know: Scientific Epistemology and Narrative in Victorian England*. Chicago/London: University of Chicago Press.

Livingstone, David N. 2003. *Putting Science in its Place: Geographies of Scientific Knowledge*. Chicago/London: University of Chicago Press.

Lorimer, Jamie. 2015. *Wildlife in the Anthropocene: Conservation after Nature*. Minneapolis: University of Minnesota Press.

Lucier, Paul. 2009. The Professional and the Scientist in Nineteenth Century America. *Isis* 100: 699–732.

MacKenzie, John M. 1988. *The Empire of Nature. Hunting, Conservation and British Imperialism*. Manchester: Manchester University Press.

Mayr, Ernst. 1986. In Memoriam: Jean (Theodore) Delacour. *The Auk* 103: 603–605.

Matthews, Geoffrey. 1980. Obituaries: Miss Phyllis Barclay-Smith. *Bulletin of the International Waterfowl Research Bureau* 46: 5.

Oreskes, Naomi. 1996. Objectivity of Heroism? On the Invisibility of Women in Science. *Osiris* 11: 87–113.

Paul, Herman. 2011. Performing History: How Historical Scholarship is Shaped by Epistemic Virtues. *History and Theory* 50: 1–19.

Pelzers, Elio. 1994. *Geschiedenis van de Nederlandse commissie voor internationale natuurbescherming, de Stichting tot internationale natuurbescherming en het Office international pour la protection de la nature*. Amsterdam: Nederlandse commissie voor internationale natuurbescherming.

Proctor, Robert. 1991. *Value-free Science? Purity and Power in Modern Knowledge*. Cambridge: Cambridge University Press.

Rayward, Boyd W., ed. 2008. *European Modernism and the Information Society: Informing the Present, Understanding the Past*. London: Ashgate.

Savoia, Paolo. 2015. Seeing and Hearing: Charcot, Freud and the Objectivity of Hysteria. In *Objectivity in Science*, ed. Flavia Padovani, Alan Richardson, and Jonathan Y. Tsou, 123–144. Dordrecht: Springer.

Shapin, Steven. 1994. *A Social History of Truth: Civility and Science in Seventeenth-Century England*. Chicago/London: University of Chicago Press.
Strasser, Bruno. 2012. Collecting Nature: Practices, Styles and Narratives. *Osiris* 17: 303–340.
Strom, Sharon Hartman. 1992. *Beyond the Typewriter: Gender, Class, and the Origins of the Modern American Office Work, 1900–1930*. Urbana and Chicago: University of Illinois Press.
Woebse, Anna-Katharina. 2011. *Weltnaturschutz: Umweltdiplomatie in Völkerbund und Vereinten Nationen 1920–1950*. Frankfurt am Main: Campus Verlag.
Zagzebski, Linda Trinkaus. 1996. *Virtues of the Mind: An Inquiry into the Nature of Virtue and the Ethical Foundations of Knowledge*. Cambridge: Cambridge University Press.

# "The Lonely Form Dies": How Epistemic Virtues Connect Roman Jakobson's New Science of Language and His Personality

**Bart Karstens**

**Abstract** Roman Jakobson (1896–1982) counts as one of the most influential linguists of the twentieth century. In the 1920s he was among the founders of structural linguistics, or shorthand 'structuralism'. With this approach Jakobson wanted to establish a new scientific standard. He assembled a number of concepts and methods stemming from a variety of disciplines (including literary studies, chemistry, biology, psychology and musicology), fused them into a more or less coherent approach and brought this approach to bear on the study of language. I will argue that this versatility can be understood with reference to a number of character traits that marked Jakobson's personality. Interestingly, these 'personal' virtues corresponded well with the epistemic virtues that were promoted in the programme of structuralism. The analysis presented here contributes to the growing body of evidence in recent historical scholarship that these two realms cannot be strictly separated. It has been noted earlier that in the case of Jakobson 'the person and the work are one'. Yet, making this connection explicit in the language of virtues represents a novel approach to the study of Roman Jakobson and his scholarship. Moreover self, science *and* society can in this case be connected through the prism of virtues, as the concept of structuralism was explicitly minted to capture the spirit of the time, both in science and beyond.

**Keywords** Epistemic virtues • Roman Jakobson • Structuralism • History of linguistics • Discipline formation • Hybridization • Anti-positivism • Holism

---

B. Karstens (✉)
University of Amsterdam, Amsterdam, The Netherlands
e-mail: B.Karstens@uva.nl



J. van Dongen, H. Paul (eds.), *Epistemic Virtues in the Sciences and the Humanities*, Boston Studies in the Philosophy and History of Science 321,
https://doi.org/10.1007/978-3-319-48893-6_10

## Introduction

Science can be understood as a pursuit of virtues. That is, the demands we place on our theories of the world can be expressed in terms of virtues such as empirical adequacy, predictive success, accuracy, precision, consistency, coherence, simplicity, fruitfulness, explanatory strength, degree of integration/unification, scope of application and problem-solving capacity. In applying this perspective to the study of past science a heterogeneous approach should be adopted. Not all of the virtues listed above need to be pursued at the same time *and* a particular virtue can be ranked higher in comparison to others.

Evolutionary theory, for example, scores well on explanatory strength as concepts of 'fitness', 'adaptation' and 'struggle for life' explain why and how species (have) survive(d). But evolutionary theory does not score well on predictive success, as new biological variants are hard to predict. In astronomy it is the other way around. This field has always scored well on predictive success but not on explanatory strength. The future position of heavenly bodies can be exactly predicted through a method of calculation, but this method can be used without being able to explain the causes of movement. These examples show that differences between academic disciplines can be carved out as differences in virtue pursuit and/or differences in virtue hierarchy. Virtues can also go contested within a discipline, as successive research programmes can be contrasted through changes in virtue preferences.[1]

Historians of science tend to identify the set of virtues in a loose manner as 'epistemic'. In philosophy of science it is however common to make a sharp distinction between epistemic virtues and other theoretical virtues. Epistemic virtues, such as consistency and empirical adequacy, are indicative of truth and therefore offer a firm basis for theory choice. Other theoretical virtues, such as simplicity, only offer further indications for theory choice (Van Fraassen 1980). In this view, the non-epistemic theoretical virtues are pragmatic criteria that function as tiebreakers when no other grounds for theory choice are decisive. They also function as pragmatic criteria in the sense that they characterize the convenient use of a theory. For example, simple theories are more convenient than complex theories because they can be more easily tested.

This philosophical distinction only makes sense when knowledge is defined in relation to truth, as in the classic definition of knowledge as 'justified true belief'. But there is no compelling reason to do so. Kuhn has famously argued that what ultimately matters in scientific paradigms is not truth but problem-solving effectiveness. It follows that both knowledge and scientific progress can be understood directly in terms of virtues, without invoking a concept of truth (Kuhn 1962; see also Laudan 1977).

[1] This also applies to whole-scale changes in thinking about science. Daston and Galison 2007 argue that the concept of objectivity has been differently interpreted over time and these changes in interpretation involve changes in virtue preferences.

I feel that this is the most fruitful way of looking at things from which, in my opinion, a number of important considerations follow. When standards of theory acceptance and measure of progress are expressed in terms of epistemic virtues, the pursuit of these virtues become primary to the endeavour we call science. Since virtues are not entities in themselves, but rather goals or 'goods' to strive for, understanding the *process* of attaining them should stand central in the study of past science. Given that 'epistemic' means 'relating to knowledge', we can consequently label all virtues that play a relevant role in the pursuit of knowledge 'epistemic'.

This includes the so-called personal virtues. Habits and character traits of human beings embody qualities, like tenacity, curiosity, creativity and honesty, which are considered essential for being a 'good', 'reliable' or 'professional' scholar (DePaul and Zagzebski 2003; Paul 2014).[2] There is a growing body of publications in support of the idea that these personal virtues relate to the more traditional set of epistemic virtues in intricate ways and are in fact hard to separate from them. The ethos of a scholar is directly reflected in the epistemic virtues that this scholar promotes. For this reason Daston and Galison propose to fuse the self and the epistemic into one ethico-epistemic analysis of past science (Daston and Galison 2008). Creyghton et al. (2016) present three case studies of 19th and early 20th century historians for whom the pursuit of epistemic, moral and political goods were closely intertwined. Other such examples are given in Shapin 2008, Tai and Van Dongen 2016 and Paul 2017.

The inclusive approach presents itself as highly useful with respect to the study of the linguist Roman Jakobson (1896–1982) and the genesis of structuralism. Jakobson connected a host of ideas and methods from a variety of artistic and scientific fields, which were fused into a more or less coherent approach to the study of language that he called structuralism. While aspects of the spheres of influence on structuralism have been intensively studied, for example with respect to phenomenology in German philosophy (Holenstein 1975) or with respect to the formalist movement in literary studies (Gasparov 2014), grasping the staggering scholarly versatility of Jakobson, and the impact this had on structuralism in its entirety has so far seemed an elusive prospect.[3]

Studying science in terms of virtue pursuit offers a solution to this problem in three ways. First, Jakobson selected ideas, concepts and methods drawn from a variety of sources that promoted the *same* kind of virtues, which included purification, abstraction, generalization and unification. Secondly, the very virtue of integration characterizes the core of Jakobson's scientific work and general attitude in life. This attitude blended in with a number of personal character traits, such as open-mindedness, broadness, creativity, erudition and generosity.[4]

---

[2] Wishful thinking, dogmatism, etc. sort the opposite effect.

[3] Testimonies of the breadth of interests and depth of mastery of subjects Jakobson possessed can be found in Halle 1983b.

[4] Clearly the virtue approach to the study of past science itself also promotes the virtue of integration, but it would go too far to claim that it is a structuralist approach. Nor can it be required that the virtue historian need to be as open-minded and erudite as Jakobson was, although these surely are lofty ideals!

Thirdly, seeking integration was also typical of the *Zeitgeist* in which many attempts to deliver an all-encompassing synthesis could be witnessed. The term structuralism was minted to capture a wholesale change in science towards anti-positivism and relationalism. Structuralism appeared to present a solution to the long-standing problem of how to anchor modern science in culture. Relationalism, and the identification of invariant properties, offered a basis for projects of social reform. The new science thus immediately led to civic engagement, in service of the virtues of solidarity and liberty. One should be wary of invoking *Zeitgeist* in historical explanation as an abstract entity that in a mysterious way exerts its influence. But when the spirit of time can be made concrete in terms of virtue pursuit, like the things people have aspired to, this yields possibilities for meaningful explanation.

In the first section of the paper I will provide a brief outline of linguistic structuralism. The Prague Linguistic Circle in many respects used Saussure's *Cours* as a point of departure.[5] Saussure had introduced important analytical distinctions such as synchrony (current operation of language) vs. diachrony (historical development of language) and *langue* (systematic rules and conventions of a linguistic system) vs. *parole* (actual language use). Jakobson continued to work with these oppositions, but at the same time tried to resolve them through an investigation of the structural relations of language. He connected this relationalism to an anti-positivist conception of science. In section two I show how this approach can be contrasted with nineteenth century linguistics in terms of differences in virtue pursuit. Jakobson drew many parallels of the new science of linguistics with other fields of study. The similarities he saw can be explained with reference to communality in the pursuit of virtues. Moreover the general holistic attitude itself promoted virtues of integration and universalism.

In the third section the early genesis of structuralist ideas is placed in the context of Russian avant-garde art in the 1910s and concomitant calls for social reform. Rethinking basic structures of society, culture, art and language were strongly intertwined and occurred to participants as aspects of the same endeavour. This has certainly contributed to Jakobson's holistic attitude, both within linguistics and in drawing connections between linguistics and other fields. In section four I show that this holistic thinking was a sign of the times, as many people sought an all-encompassing synthesis to connect science, society, culture and art. In this section I also seek to explain the holistic attitude in terms of personal virtues that marked Jakobson's character and working ethos. In closing I briefly consider whether the combination between personal virtues, relationalism, anti-positivism and a strong preference for virtues like generalization and simplicity form an inevitable bond.

[5] The *Cours de linguistique générale* is a posthumously published book based on lecture notes by students of Saussure. It should not be read as a finished theory of language but as a collection of groundbreaking ideas that require further development by others.

## The Structuralist Approach to the Study of Language

Roman Jakobson enrolled at the historico-philological faculty of the University of Moscow in 1914. Already one year later he was among the founders of the Moscow Linguistic Circle. The aim of the Circle was to advance the study of linguistics, poetics, metrics and folklore. Together with the OPOJAZ (Petersburg), which translates as "Society of the Study of Poetic Language", the Moscow Circle represented the leading movement of literary criticism in Russia from the 1910s to the 1930s. In 1920 Jakobson pursued his doctoral studies in Prague, where he initially arrived as a member of a Soviet diplomatic mission. He stayed in the Czech republic for almost two decades. Together with Mathesius, Trubetzkoy, Makarovsky and Wellek (among others) he formed the Prague Circle, which was to become the cradle of linguistic structuralism.

At the Masaryk University in Brno Jakobson was appointed professor in 1933. There he developed his seminal theory of distinctive features in the field of phonology. After the rise of the Hitler regime Jakobson fled the Czech Republic and eventually landed in New York after traversing Germany and Scandinavia. In the United States he became a professor at Harvard from 1949 until his retirement. His ideas have influenced numerous scholars in the U.S., and abroad, most prominently Claude Lévi-Strauss and Noam Chomsky.

Following Saussure, the Prague Circle put central emphasis on the study of language as a form of social behaviour. Language was first and foremost a communicative tool through which signals can be transmitted, with the goal to provide information to others, to establish social relations, to give vent to emotions and/or to create art. A key element of structuralism was 'the sign'.[6] A sign, in Saussurian terms, is made up of a signifier (*signifiant*), for example the shape of a word, and of the concept (*signifié*) that is signified. Saussure held that the relation between *signifiant* and *signifié* was arbitrary, that is, any word could in principle stand for any concept. In structuralism on the other hand, a signifier *cannot* be fully distinguished from its signifying properties. A typical structuralist assumption was the correspondence of form and meaning. According to structuralists the form of, for example a word, carries meaning in view of the place it occupies in language. The very appearance of a form already signals a difference with other forms *and* the recurrent and systematic appearance of a form in linguistic expressions shows how it is related to other forms. To a large extent this determines the meaning of the sign.[7]

A linguistic sign could be defined as a formal and recurring property of a language. From this it was possible to extract a set of invariant properties, which together represented the basic building blocks of a language. A first major objective of structuralism was to establish what the invariant properties of a language were.

---

[6] The sign was not a new idea, but gained a central place in linguistic theory with structuralism. See Jakobson's *longue durée* history of 'the sign principle' in Jakobson 1971a.

[7] Comprehensive overviews of Jakobson's structuralism are given in Waugh 1976 and Sangster 1982.

This was a difficult task because variants had to be fitted into invariants, which was not a straightforward thing to do. A further complication was that invariants had to be distinguished on a variety of levels (phonology, morphology, syntax, etc.). This work also proved to be entangled with the psychological dimension of language use. The process of disentanglement of all these levels was thus difficult, but, according to Hunt, in the end potentially very rewarding: "Decisions about how many variants can be fitted into a specific invariant have raised many important questions about degrees of abstraction, about psychological reality, about the nature of linguistic perception, about the relationships between grammar and the lexicon-in short about all fundamental questions of language and linguistics" (Hunt 1983: 56).

The assumption of a basic form-meaning correspondence did not yet provide a complete specification of the meaning of linguistic signs. The full meaning of a linguistic expression only came about through the relations of this expression with other linguistic expressions. The second main objective of structuralism was to provide an account of the relations between invariant elements of a language. This set provided the basic relations of a language and together they constitute its structure. The central idea now was that all linguistic behaviour (*parole*) could be explained with reference to this fundamental and 'deeper' structure.

A language can only function as a communicative system, when basic relations are repetitive and hence recognizable for the speakers of the language. At all stages of development a language must thus have a structure of relations between invariant properties, as otherwise it would be unintelligible. Jakobson expressed this holistic perspective of relative invariance as follows: "A word form is perceptible to us only due to its recurrences in the system of language. The lonely form dies" (Jakobson 1979: 336). It is important to note that members of the Prague Circle did not perceive the stability provided by the language structure as static. It was rather seen as a dynamic equilibrium (Galan 1985: 82–140). Signs can evolve through changes in the relation between *signifiant* and *signifié*, and as a consequence basic relations can change and alter the structure of the language. While relations between form, concept and reference are certainly not arbitrary in structuralism, Jakobson recognized that users of language realize that there is a difference between them. This, according to him, is even a condition for dynamic language use: "[S]uch antinomy is vital because without contradiction there is neither movement of concepts nor movement of signs" (quoted ibid.: 109). This dual nature of an identity, that is a contrast at the same time, is typical of Jakobson's and other structuralist attempts to overcome oppositions (such as *langue* vs. *parole*) while not dissolving them. Change always happens against a relatively stable background, provided by the structure of the language. The structure provides the framework in which meaningful variation is possible.[8]

The clearest demonstration of the structuralist approach to the study of language is in the area of phonology, the subfield of linguistics that is concerned with the

[8] This idea was behind the project of optimization of language. See section 3 below on the projects on societal and linguistic reform, which had inspired the young Jakobson.

systematic organization of sounds in languages.[9] A phoneme is the minimal unit in a language that can serve the purpose of differentiating meaning. When combined phonemes make up words. Phonemes are not letters however, but basic sounds, like /z/, which occurs in 'zoo', but also in 'nose'. Jakobson's groundbreaking insight was to interpret phonemes as 'bundles of sound'. In his theory the phonemes are *not* the basic elements of a language, but they are constituted by sets of *distinctive features*. Examples of such features are onset, voicing, laterality, nasality, etc. Contrasts between phonemes can be described in terms of differences between features, just like molecules can be distinguished through differences in constituting atoms.[10] This compositionality extends upwards as words can be interpreted as compounds of phonemes, sentences as compounds of words and texts as compounds of sentences.

As just explained, the point of structuralism is that associations between features, phonemes, words, etc., do not occur arbitrarily but in systematic ways. That is, they have to respect basic structural properties, which are expressed in terms of binary *oppositional* relations. In Jakobson's theory of phonology each distinctive feature involves a choice between an opposition with a marked and an unmarked side. Voicing (sonorous sound) contrasts the phonemes /b/ (voiced) and /p/ (unvoiced). But a similar opposition is found between /d/ (voiced) and /t/ (unvoiced). This shows that such oppositions are recurring in language. Other differences between the phonemes in the example are given with reference to other features. An optimal analysis should yield the least number of feature specifications per phoneme. The combination of features follows conventions that are specific for each language and embodied in it. The total number of phonological oppositions, the hierarchy between them and the conventions for combining them need to be discovered by the linguist.[11] This asked for a conquering mentality, which befitted Jakobson, as he was by temperament a *conquistador* and not a doubter (Gardner 1985: 202, 236). Taken together the set of binary oppositions constitutes a system of relations. In his final theory Jakobson distinguished 13 distinctive phonological features, which were further classified into three main categories: sonority, propensity and tonality (Jakobson and Halle 1956).

According to the members of the Prague Circle the structure of languages consisted of oppositions because humans tend to perceive things in terms of oppositions. Properties of the human mind not only determine the way language operates, but also the physical reality of language. Distinctive features are both physically and

[9] The rise of phonology to a well-respected subdiscipline of linguistics has much to owe to the work of the Prague Circle.

[10] For the influence of ideas from chemistry on early structuralism, see Sériot 2014.

[11] Phonology served as a model for structuralist analysis in other areas of language. With respect to the study of word order (syntax) Jakobson applied the concept of opposition in his analysis of the Russian case system and categories of the verb. In another application of the 'phonological' approach, Jakobson developed a theory of maximal contrast in the sequence of sound development in children, see Jakobson 1941.

psychologically real because they have evolved for specific purposes.[12] In this respect we witness a strong influence of German phenomenology and psychology on early structuralism. Carl Stumpf, a prominent phenomenologist, put forward an analysis of speech sounds that served as an important source of inspiration for Jakobson's phonology.[13]

Structuralist linguistics follows the reductionist strategy of explaining the variety of language use with reference to a restricted set of invariant linguistic elements and their relations, both of which obtain on a deeper structural level. This involves selection, generalization and simplification. The compass of conquering the invariants was thus set by the pursuit of the virtues of economy, generality and simplicity (Chvany 1984: 43). A holistic attitude was also strongly present.[14] We find it in the assumption of meaning-form correspondence, in the notion that a signifier signifies something already in view of mere otherness, in the idea that the full meaning of a linguistic expression can only be established in relation to other linguistic expressions and that the functioning of language as a whole can only be understood when the structure of systematic relations is understood. Jakobson was convinced that this holistic attitude and the pursuit of virtues in structuralist linguistics fitted a new conception of science, which is the topic of the next section. They also strongly intersected with avant-garde art and civic engagement, which are discussed in the last two sections of the paper.

## Structuralism as a New Science

Saussure had claimed autonomy for linguistics because according to him the systematic aspects of languages constituted a unique object of research. Following up on this, Jakobson asserted that with structuralism linguistics had become a real independent science.[15] Saussure and Jakobson were not the first to claim to study language 'in and for itself'. In the mid-nineteenth century August Schleicher had proposed to separate linguistics entirely from philology and turn it into full natural science, which he proposed to call 'Glottik' in analogy with 'Physik' and 'Botanik'. This was a radicalization of historical and comparative linguistics that in the early nineteenth century had started to conceive languages as organisms and had in part been modelled on methods of comparative anatomy (Karstens 2012). Schleicher took the organism metaphor to extremes and argued that the life of languages

[12] On the relation between evolution and language in structuralism see next section.

[13] Toman 1994: 107. Holenstein 1975 simply calls structuralism phenomenological. Flack 2016 however argues that the influence of German phenomenology on structuralism has not been adequately studied yet and needs to be reassessed.

[14] Another branch of psychology namely *Gestalt* psychology has influenced the anti-atomistic and 'Ganzheitliche Einstellung' among structuralist linguistics according to Holenstein 1975.

[15] For contemporary reflections on the autonomy of linguistics, see Lazard 2012 and Givon 2016.

followed strict laws of biological development (*Lebensgesetze*) and should be studied with the same rigor as in the natural sciences.

A strange consequence of the 'Schleicherian paradigm' was that changes in language over time were studied irrespective of the behaviour of speakers of the language (Koerner 1982). Especially the Leipzig-based Neogrammarian school, which directly preceded Saussure, distanced itself from Schleicher's approach and claimed a role for psychology in the explanation of language change because psychology could provide insight in the way individuals bring this change about. Yet in the Neogrammarian approach psychological explanation was no more than a supplement to the main historical and comparative methodology. Also they considered psychology to be an empirical science that produced lawlike regularities.[16]

For these reasons the Neogrammarian school was the object of attack in Jakobson's writings. While he applauded the incorporation of psychology in linguistics he denounced the empiricist positivism of the Neogrammarians.[17] For Jakobson linguistic science should not be about collecting data and discovering laws, but about discovering structural properties. In his view it was the search for the basic structural relations of a language *and* their communicative and creative functions what makes a purely linguistic science. A relationalist and anti-positivist attitude paved the way for a new scientific ethos.

We can understand the shift from the positivistic historical and comparative linguistics to the relationalism of structuralism in terms of a shift in virtue preferences. The former can be associated with the pursuit of the virtue of empirical adequacy coupled with a low tolerance for irregularities. The latter was guided by the pursuit of generality, simplicity and economy.[18] The tolerance for deviations significantly increased as a consequence. For Jakobson, reducing linguistic phenomena to invariant properties of a language did not mean that everything had to be covered. The trick is to fuse the most linguistic data, with the least amount of 'noise', into a framework with a restricted number of structural features.[19] Simplification allowed for exceptions to the rule because languages are alive, and it is in their nature to 'wander about' and change. It followed that at every level of linguistic inquiry exceptions are tolerable.[20] According to Stankiewicz structuralism was more a method of discovery than a theoretical straightjacket (Stankiewicz 1983: 26). Still, structuralist research is guided by the idea that one will find oppositional relations. In this sense 'theory' steers the discovery process, which yielded another point of departure from positivism.

All this did not mean that structuralist linguists were averse to lawlike description, but rather that they demanded a deeper explanation of the obtained regularities,

---

[16] For an extensive description of the Neogrammarian school see Morpurgo Davies 1998.

[17] He also thought that the psychology of Herbart, Lazarus and Steinthal they relied on was old-fashioned and needed to be replaced with ideas from phenomenology and *Gestalt* psychology.

[18] I mean generalization and simplification as embedded in the holistic approach and *not* in the sense of empiricist inductivism.

[19] For philosophical reflection on identifying patterns in the world see McAllister 2011.

[20] As pointed out in Hunt 1983 and Chvany 1984.

related to the systematic properties of a language or to the structure as a whole. At the end of his life Jakobson expressed this as follows: "Since my first steps in phonology I have been continuously attracted to the search for the ultimate constituents of language and the powerful structural laws of the network they comprise" (Jakobson 1985b: 372).[21] In another citation he explained this view on laws in more detail:

> Only functional analysis of linguistic facts, only a systemic approach to language change, one that considers language change from the perspective of a system undergoing change, makes it possible to construct a theory of evolution of language and raise the history of language from the level of episodic disciplines to the rank of system-oriented science. Only this approach can create preconditions for the discovery of basic structural laws of language; the understanding of sound laws receives a new, sharper meaning, totally new perspectives open up for the comparative method (quoted in Toman 1994: 69).

There was thus even a place for the historical and comparative method in the structuralist framework, but its role had become secondary. Something similar happened with genetic classification. Typologies in this approach were based on evolutionary trees, with splits in families of languages, such as the Indo Germanic family, the Semitic family or the Finno Ugric family. Jakobson advocated a different typology based on the structural characteristics of languages.[22] It was still possible to find genetically related groups in this typology but this had become of secondary importance with the identification of structural similarities as the new overarching method. Structuralists replaced the concept of *Sprachfamilie* with that of *Sprachbund*, the key example of which is the so-called Eurasian *Sprachbund* which roughly corresponded with the boundaries of the Russian Empire.

The new science of linguistics was linked to a major change in scientific thinking that was happening at the time, at least according to Jakobson. In this context he used the concept of structuralism for the very first time:

> Were we to comprise the leading idea of present-day science in its most various manifestations, we could hardly find a more appropriate designation than structuralism. Any set of phenomena examined by contemporary science is treated not as a mechanical agglomeration but as a structural whole, and the basic task is to reveal the inner, whether static or developmental, laws of this system. What appears to be the focus of scientific preoccupations is no longer the outer stimulus, but the internal premises of development; now the mechanical conception of processes yields to the question of their functions (Jakobson 1929: 11).[23]

This was not just an incidental observation as Jakobson devoted a number of papers to demonstrate parallels with other fields. One of the most striking of these was with Einstein's 'new' physics.[24] For a number of reasons Jakobson felt attracted to the

[21] See also Part 2C of his *Selected Writings*, vol. 7, which is titled *Towards a Nomothetic Science of Language*.

[22] A key text in this respect is Jakobson 1931.

[23] Percival 2011: 244 gives the same quote and suggests that this is actually the earliest use of the word structuralism by a linguist.

[24] See Jakobson 1985a and 1985b as well as the remarks in Stankiewicz 1983: 21–22 and Ivanov 1983: 47–48.

theory of relativity. First, Newton's law of gravity had turned out not to be the final word in physics. Einstein had not only shown that this law could only adequately describe a particular set of phenomena, but more importantly, he had offered a deeper explanation of gravity through the curvature of spacetime, which was essentially a relationalist concept. This fell in line with the structuralist attitude towards laws that did not deny that lawlike regularities could be found in linguistics but always demanded a deeper explanation of them to justify their validity.

Also Jakobson liked the very idea of the unity of spacetime. He thought that space could be equated with the synchronic and time with the diachronic and hence the concept of spacetime offered possibilities to overcome the dichotomy between synchronic and diachronic study of language, arguably one of the thorniest issues in all of linguistics. But Jakobson has not clarified how this analogy had to be further worked out in detail. Other concepts Jakobson was attracted to were symmetry and invariance. Especially the (mathematical) concept of invariance was a cornerstone of the structuralist method, as we have seen above. Jakobson even went as far as to suggest that the concept of relativism was transmitted to Einstein via his teacher Winteler, who according to Jakobson had developed a kind of pre-structuralist linguistics that deeply influenced Einstein. Would this claim be true, it would of course greatly enhance the status of linguistics and the role it had to play in the new relationalist conception of science. It is however more likely that Jakobson 'forced' bits of evidence into a rather blown up story (as demonstrated in Kohrt 1984).

The similarities between Einstein and Jakobson are also striking in terms of virtue pursuit. According to Van Dongen (2010) generalization, in terms of looking for invariant structures, logical simplicity, in the search for the mathematically most natural structures, unification, in the search for unified field theory, and a preference for deduction over induction, all determined Einstein's scientific attitude, especially in the later part of his career. As we have seen above, the virtues of generalization, simplicity, unification coupled with anti-positivism all marked the programme of linguistic structuralism too.

Next to physics, Jakobson also drew parallels to other fields that reinforced the conviction that he was on the right track. After a lecture by the musicologist Becking in Prague, Jakobson wrote a contribution for the *Prager Presse* in 1932 titled *Musikwissenschaft und Linguistik* in which he noted many similarities between the two fields (Karstens 2017). Next to this he supported Berg's anomalous interpretation of evolutionary theory as published in his book *Nomogenesis* (1922) (Toman 1994: 123). In this interpretation limits are set on the possible variation of species and an important role is played by the notion of convergence. This idea can be found in structuralism too because variation in expression is bound by the structure of the language, and in this sense limited. Finally, Jakobson drew a parallel between the method of structuralism and the method of vivisection. He argued that because it is extremely difficult to transpose oneself into the past, the reconstruction of the past is inevitably deformed. Only linguistic facts of the present can be experienced without distortion. Hence the idea that scientific vivisection would be potentially more fruitful than the dissection of 'dead corpses', that had been the method of historical and comparative linguistics.

In a retrospect, Jakobson described his career as a "lifelong tendency to overcome in scholarship mere short-sighted empiricism on the one hand and abstract speculative dogmatism on the other" (Jakobson 1985b: 377). The new science that Jakobson so strongly advocated can be explained as a shift from mechanical objectivity to structural objectivity following Daston and Galison 2007. In mechanical objectivity the highest goal is to reproduce nature 'as it is'. The role of the scientists is essentially unimportant, he or she is a mere 'worker' in service of reproducing the correct image of nature. The notion of structural objectivity shifts the objective image of nature from the registration of surface phenomena towards deeper structures that can be used to *explain* the occurrence of surface phenomena. Structural linguistics must perhaps be considered as the first science that can be genuinely said to pursue a programme of structural objectivity.[25]

At first the role of the individual scientist in the discovery of invariant structures is still essentially unimportant. Yet, with the rise of a new 'ethic' that Daston and Galison call 'trained judgement' this starts to change. Trained judgment is all about the identification of patterns. On the one hand this involves active and highly trained experts, but on the other hand the ethic of trained judgement recognizes that intuition and creativity play an important role in science as well because in the discovery of patterns processes of abstraction, selection and drawing similarity relations lean on intuitive and creative functions of the mind. The discovering mentality requires confidence in one's ability to judge. The notion of 'trained judgement' accords well with Jakobson's work in discovering linguistic structures and the *conquistador* temperament that he had. While Daston and Galison tend to distinguish structural objectivity from trained judgement one could in the case of Jakobson also argue that aspects of both jointly characterize his scientific attitude.[26]

Jakobson used to say that because he was a linguist nothing pertaining to language was alien to him: "Linguista sum, linguistici nihil a me alienum puto." This was an appropriation of the saying "Homo sum, humani nihil a me alienum puto" which is attributed to Publius Terentius Afer who lived in the second century B.C. The implication of this dictum was that one should keep an open mind to all areas in which language played a role, and which could have an impact on the study of language. We have seen above that he recognized the pursuit of similar ends in physics, musicology and biology, even though in the case of both physics and biology this had a rather forced character. Next to this it is also clear that Jakobson took over ideas such as *Gestalt*, from psychology, 'invariance' from mathematics, 'limited variation', from biology and the idea of compound structures from chemistry. All of these gained a prominent role within structuralism.

---

[25] Daston and Galison 2007: 255–256 briefly discuss the rise of the concept of structure in scientific discourse. The birth of structuralism is only mentioned in a footnote and incorrectly depicted as being antihistorical.

[26] Interestingly, given the parallel that has already been drawn between Jakobson and Einstein, Daston and Galison 2007: 305 inquire whether Einstein can be considered as a typical structural objectivist, to which they answer yes *and* no. Like Jakobson, Einstein too cannot be fitted neatly in either of the two categories. See in this respect also Van Dongen 2017.

I conjecture that ideas from other fields were selected through an assessment of the epistemic virtues they helped to promote. The ideas from psychology and chemistry helped to consider the whole in its relation to its constituting parts and hence they promoted the virtues of integration and unification. Ideas such as 'limited variation' and 'invariance' serve generalization and simplification as they help to reduce the set of directly observable linguistic phenomena to the basic underlying structure.

## Avant-Garde Art and the Genesis of Structuralism

During his years in Moscow in the 1910s Jakobson had frequent contact with literary theorists, poets and painters of the avant-garde movement. Especially poetry had a major effect on his thinking about language.[27] Jakobson was attracted to the peculiar double status of poetry as it is at once made of language and *about* language and he attributed high value to the self-reflective function of poetry. Through this function, poetry could act as a path-breaking laboratory for insights into the structure and function of the sign.

With poetry, Jakobson had formalist poetry in mind. The goal of formalist poetry was to analyse the content of linguistic forms, especially how they were constituted. As this also involved a reflection on the meaning of linguistic expressions, formalism went to the heart of the study of the sign, that is the correspondence between form and meaning. Jakobson was most impressed by the work of the poet Khlebnikov (see esp. Jakobson 1979).[28] Khlebnikov perceived language as a limitless and boundless playground of interconnections between form and meaning. Because we cannot see the interconnectedness of all meanings due to our cognitive limitations, the task of the poet, for Khlebnikov was to show latent potential in language by laying bare missing links in the great chain of meaning. He 'played with words' and through endlessly varying constituent components he was able to create new words in Russian.

This play with words was not a random exercise but was based on the idea that a language is an integral system that consists of sets of binary and minimal oppositions. The minimal pairs set limits to the possible sound combinations out of which words can be created. Khlebnikov's word creation technique thus stood in close proximity to structuralist linguistics, as its starting point was given by invariant

[27] Key texts are Jakobson 1981a and Jakobson 1981b. Secondary literature with special focus on Jakobson and poetry includes Birus, Donat and Meyer-Sickendiek 2003, Kiparsky 1983 and Rudy 1978.

[28] Steiner 1984 makes clear that formalism was not a unified movement and that there were as many formalisms as there were formalists. He also demonstrates that Khlebnikov was very influential in the Moscow Linguistic Circle, which is why we focus exclusively on him in this paper.

properties of the language.[29] Knowledge of the basic structural relations could show unused areas of expressing meaning and help to determine the scope of meaningful variation. An important device in this respect was also given by the method of *ostranenie*, or defamiliarization.[30] Trying to defamiliarize oneself from what is 'known' reveals the essence of a thing, and then consequently makes possibilities for variation visible (Gasparov 2016). According to Gasparov (2014), Khlebnikov's poetry reads as if it was deliberately written in support of structuralist linguistics, which was of course not existent yet. However, it is more prudent to say that Jakobson took over the key distinction between a deep and a surface level of language as well as the regular recurrence of linguistic patterns from formalism. This regularity is something Jakobson appreciated in all of the arts: "There is a mathematics of the word (as there is a mathematics of all the other arts), especially in poetry" (Jakobson 1981b: 95).

The formalists did not just play with words for fun. Their project of language reform had the goal to maximize freedom of expression. From this a direct line runs to projects of social reform, because language reform is about individual freedom and exerting control over communication. A similar combination of both reform projects was present in Italian futurism. Yet, the Russians developed their own approach to language reform in contradistinction to the Italians. The difference became apparent when Filippo Marinetti, the author of the 'Futurist Manifesto' (1909), paid a visit to Moscow in 1914. In a lecture Marinetti proposed to liberate speech by freeing words from the bonds of grammatical rules and through onomatopoetic and emotive interjections create expressions that appeal directly to the senses (Gasparov 2014: 87–88).[31]

To expand and renovate vocabulary in this way, and to let go off all syntactic prescriptions appeared to the Russians as shallow and naïve. Sacrificing all convention made innovation unintelligible, or a purely individual matter. For the Russians, Italian futurism downplayed what was in effect the most important aspect of language: its social and hence its signifying functions. It is the mastery of structural concepts, which frees the language speaker, *not* giving up on all rules. In this respect Jakobson approvingly quoted the composer Stravinsky who had said that "renewal is fruitful only when it goes hand in hand with tradition" (Jakobson 1985a, b: 186). Linguistic reform should thus start from collective memory and collective values. Gasparov captured the Russian spirit as follows:

---

[29] Galan (1985) shows that structuralism eventually deviated from formalism in important respects. Nonetheless the basic idea of structural oppositions setting limits to possible variation is present in both and Jakobson repeatedly stressed that linguists 'learned from the poets' when it came to the formulation of structuralism.

[30] This concept stems from an essay by formalist literary theorist Viktor Shklovsky: 'Art as Technique' (1916), also sometimes translated as 'Art as Device'.

[31] Such ideas about language reform were not uncommon at the time. Both Dada and surrealist movements promoted spontaneous language use, freed from rules and 'formulas', see Van Spaendonck 1977.

> Any advance into the future proceeded by evoking memories of the past; any individual departure from convention relied on the collective inheritance of language memory. .. True innovation set in motion whole layers of collective language consciousness in order to renew and reshape it; the deeper into the past those layers had reached the more powerful would be the effect of their futurist transfiguration (Gasparov 2014: 89).

Kiparsky has drawn an explicit connection between *this type* of futurism, avant-garde art and a general reorientation in science:

> It was a period when aesthetic as well as scientific values were undergoing a radical reorientation. Nineteenth-century realism in art and literature had been dominated by the same concern for fullness of descriptive detail as nineteenth-century linguistics, with its emphasis on phonetic and lexical problems. The question was: "How much can I put in?" Artists around the first decade of this century lost interest in this ambition and began to ask instead: "How much can I leave out?" Jakobson has recalled his excitement over his first encounter with the works of the Cubist avant-garde and credits them with having inspired his search for abstraction and structural invariants in language (Kiparsky 1983: 28).

Again we find the opposition between the epistemic virtues empirical adequacy on the one hand and generalization and/or simplicity on the other hand.

This new mindset was also visible in the work of Russian painters like Matyushin and Malevich both of whom Jakobson admired and with whom he stood in direct contact. In their work we find attempts to *overcome* limitations of direct perception. Abstraction was a way to arrive at what was essential and simultaneously to reveal new hidden dimensions. Matiushin for example stated that the mistake of Italian futurist painting was to stick to external manifestations of 'new movement' like skyscrapers, automobiles and factories. The real point of futurism for him was to capture the intrinsic forces behind the external manifestations of movement and proceed to rebuild society, culture and science on the basis of this awareness. Jakobson translated the 'intrinsic forces' to the laws that govern within a structure. He also connected what was happening in science to what was happening in art:

> Those of us who were concerned with language learned to apply the principle of relativity in linguistic operations; we were consistently drawn in this direction by the spectacular development of modern physics and by the pictorial theory and practice of cubism, where everything is based on relationship and interaction between parts and wholes, between colour and shape, between the representation and the represented. "I do not believe in things", Braque declared, "I believe in their relationship" (Jakobson 1962: 632).

Thinking of relationships in social terms means to think about the collective. By quoting the 'Futurist Manifesto' Jakobson claimed in a letter that it was the right of the poets "to stand on the rock of the word 'we' amidst a sea of boos and outrage." And in the same letter he also wrote about linguistics: "the strength of our science rested precisely on the Futurist rock of the word *we*."[32] The 'we' concept was not confined to national boundaries.[33] We have already seen above that the classification

[32] Quotes from a 1928 letter by Jakobson, printed in Toman 1994: 47.

[33] Kursell 2010 observes that formalism yielded a conception of language independent from specific national interests. According to her this allowed linguistic theories based on this conception to move across national boundaries.

of all languages based on structuralist characteristics led to the postulation of a Eurasian unity (see Toman 1994; Friedman 2014; Sériot 2014). The Eurasian *Sprachbund* set forward a supranational identity that had strong political implications, which continues to be the case to the present day.[34] Ivanov suggests that the theory of distinctive features was the product of minds, suffering from divisions in Europe, and hence seeking integration (Ivanov 1983: 49). It is however hard to verify this suggestion with a reference to one of Jakobson's papers or letters. The claim is also somewhat redundant because longing for integration was a ubiquitous phenomenon and not a virtue of the distinctive feature theory alone (see next section).

We should be careful to identify structuralism with one clear and distinct worldview (Percival 2011).[35] We have seen above that Russian futurism was different from Italian futurism, in part due to its close links to formalism. In Russia of course, a model of societal reform was radically implemented when the Bolsheviks took power in 1917. Because Soviet ideology was supranationalist and, at least ideologically, involved the liberation of the people from the yeast of traditional power, the group of Russian formalists and futurists initially welcomed the Revolution. However during Stalin's rule the doctrine of social realism was implemented and formalism was deemed incompatible with it. Formalism in this view was no more than frivolous experimentation with a subversive context. Formalist work was seen as mere play with form that had dangerously lost contact with reality. An accusation of having produced formalist art could lead to banishment to the Gulag, if not worse. Although Jakobson went to Prague as an attaché in a Soviet diplomatic mission it is for good reasons that he chose not to return to Russia afterwards. His first return visit to Moscow took place in 1956, 3 years after the death of Stalin.[36]

As late as 1950 Stalin thought the topic of linguistics important enough to devote an interview and a number of follow-up letters to the subject under the title 'Marxism and the Problems of Linguistics' which were published in *Pravda* in June 20, July 4 and August 2, 1950. One of the problems was to explain how it could be that the same language was used in the earlier capitalist system and the current socialist system. Wasn't language part of human culture and wasn't culture always dependent on the economic base structure? Stalin solved this problem by insisting on an instrumentalist view on language as a social medium, that can be used by anyone. Language was thus independent of any political system, and of any class using it, just like machines or other means of production were. Still, Russian had been improved and replenished under Soviet rule but the structure of the language, in all essentials, has remained the basis of modern Russian. This curious contribution to

[34] For a recent defence of a version of this theory see Marácz 2015.

[35] For example, structuralism is often associated with modernism. Yet Toman 1994: 86 notes that modernism was a severe anathema in Jakobson's vocabulary. For him, modernism was a style characterized by decorativism, eclecticism, symbolist aestheticism and a score of other "un-constructive" attributes.

[36] According to Ivanov 1983: 49 Jakobson remained faithful to the Russian version of Futurism until the end of his life.

the history of linguistics shows how strong the science of linguistics could in actual practice be connected to the realm of politics, just as the futurists had proclaimed, if only their call for reform differed greatly from the actual horrendous realization under communist rule.

## Longing for a Synthesis

The previous section has shown that the virtues of integration, abstraction and purification that marked the new science of linguistics resonated with the realms of art and politics. With respect to societal reform the stress on the collective and the 'we' connects to the virtues of unification and integration. With respect to art the virtues of simplicity, abstraction and purity served as categories of aesthetic judgments as well as scientific ones.[37] Next to this, both change in language and change in society had to respect limits set by its structure and its tradition. This suggests that the discussion on the role of virtues in science should not just focus on the scholarly person and categories of epistemic assessment, but also include social and artistic realms to complete the picture.

We have seen above that structuralism was permeated with holistic thinking. Holism could be seen in the assumption of meaning-form correspondence, in the notion that a signifier signifies something already in view of mere otherness, in the idea that the full meaning of a linguistic expression can only be established in relation to other linguistic expressions and that the functioning of language as a whole can only be understood when the structure of systematic relations is understood.

In philosophy holism is often opposed to positivism, atomism and reductionism. Holism cannot be squared with a form of radical empiricism that as the saying goes 'fails to see the wood for the trees'. Because holistic explanations start from a sense of what the whole is and how it functions, they tend to score well on the virtues of explanatory strength, integration and unification. Perhaps in a less explicit sense the term holism can also be applied to the general feeling that particular things hang together and we should therefore focus on the relations that bind them. In this sense too a holistic attitude is almost synonymous with the pursuit of the epistemic virtues of integration and unification.

The pursuit of these virtues is visible in the work of Jakobson in a number of ways. In the study of oppositions, such as between persistence and change, regularity and exception and synchrony and diachrony, he never declared any of these poles irrelevant or trivial, but tried to precisely determine the oppositional relation *and* indicate ways to overcome them at the same time. A language to Jakobson was first and foremost a social medium. Any project of linguistic reform had to respect the collective. This collective was broadly interpreted as an overarching geopolitical unity in connection with the Eurasian *Sprachbund* idea. Next to this Jakobson's

---

[37] The idea that aesthetic and rational judgments are complementary and parts of the same endeavour is explored in McAllister 1996.

openness towards any field of study that appeared relevant to linguistics and to structuralism in general reveals an integrative attitude.

Soon after Jakobson's death a book appeared with recollections about his personality by a substantial number of authors. From the tributary contributions a picture emerges of Jakobson's character that tunes in with holistic thinking and an integrative attitude. The authors repeatedly refer to his openness, broadness, wide range of academic interests, depth of mastery of this range, erudition, generosity, ability to see the interconnection of subjects, ability to detect unity in variety, simplicity, creativity and total absence of defensiveness. Some of these characteristics may appear to belong to the standard repertoire of tributary commemorations. Yet, if Jakobson's breadth and depth awed no lesser scholars than Claude Lévi-Strauss, Noam Chomsky and I. Bernhard Cohen, some extra significance may be attached to these descriptions. In any case, all of these character traits naturally connect to the holistic virtues of integration and unification, but also with the virtues of generality, abstraction and simplicity that are typical of structuralism.

Gasparov has interpreted the general holistic attitude in Jakobson as a neo-romantic longing for an all-encompassing synthesis, which became a formative force in shaping his views on linguistics (Gasparov 2014).[38] This longing for a synthesis is the central topic of Baneke 2008. He argues that 'synthetic thinking' as an attempt to connect, culture, nature and science was ubiquitous: "In the 1st half of the 20th century everyone appeared to be in search for a synthesis, an overarching philosophical system, in which the natural sciences, humanities, morals and the social sciences were united. It was the holy grail of that time" (Baneke 2008: 121, my translation).[39]

Two things motivated this search for an overarching synthesis. First, the increase of specialization and differentiation within the academic system led to a fragmented picture of knowledge pursuit. This prompted a desire for a perspective from which an integrated picture could be derived. Secondly, people felt that science could not answer the questions of life that really mattered to them. A clear illustration of this is the so-called bankruptcy of science debate. Inflated 19th century positivism was thought to be bankrupt because it did not help the New Man to find his place in modern society. In 1935 Dutch historian Johan Huizinga for example claimed that modern science had not yet found its place in contemporary culture. It could also be argued that the problem was that culture was lagging behind. Spengler famously warned that culture would recede into civilization if it did not react properly on the achievements won in science. Whether it was science or culture lagging behind, anchoring science in general culture was felt as one of the biggest challenges of that time. The answer was sought in a grand synthesis and structuralism appeared to provide one such synthesis.[40]

---

[38] For the connection between holism and anti-positivism in Jakobson see also Sériot 2014.

[39] See also the related discussion on 'synthetic science' in Sériot 2014.

[40] Another one was holism itself. This concept was minted in the context of evolutionary biology by Jan C. Smuts (1926) who defined holism as follows: "The tendency in nature to form wholes that are greater than the sum of the parts through creative evolution."

The prism of virtues has made connections visible between Jakobson's personality, his structuralist approach in linguistics and his general views on science and society. With so much integration at hand one begins to wonder whether everything in Jakobson' work, personality and worldview indeed fitted smoothly together. I mention two possible forms of tension. With respect to Jakobson's character his openness, generosity, broadness and holistic attitude appear to conflict with the polemical nature and conquering temperament he also possessed. Another form of tension can be seen between the claim that linguistics, with the advent of structuralism, had become and independent science and the repeated emphasis on the parallels and deep connections between linguistics and other fields of inquiry. The very motto Jakobson used all his life that nothing pertaining to language is alien to him defies the claim for linguistics as an autonomous science.

In both examples the virtue of integration comes into conflict with separating and isolating impulses. If we consider linguistics alone in this respect, there was the practical, and at the same time, daunting problem of maintaining an overview of the field while it expanded in breadth and depth. It is fair to say that linguistics has broken up into a variety of, sometimes even mutually exclusive, subdisciplines. This is certainly at odds with Jakobson's convictions and often experienced as seriously problematic, but it has proven to be difficult, if not impossible, to maintain all forms of language study under one umbrella.

## Conclusion

While much has been written about almost all aspects of Jakobson's scholarship, studying him through the prism of virtues has brought about a deeper understanding of his versatility and the impact this had on linguistic structuralism. Structuralism promoted the epistemic virtues of simplicity, generalization, purification, abstraction and unification. Structuralist linguistics was a merger of elements stemming from poetry, chemistry, mathematics, psychology, biology and phenomenology. I have argued that these elements, such as *Gestalt* from psychology, 'limited variation' from biology *and* formalism and 'invariance' from mathematics, were selected by Jakobson because they promoted the epistemic virtues of structuralism.

Yet, the structuralist approach was in the process of formation when the selection of its constituent elements took place. Collecting them could thus not have occurred with reference to the epistemic virtues of a finished theory, but must have been guided by something else. I suggest that the character of the scholar plays the most important role in this respect. Jakobson's conquering mentality agreed with the method of discovering structural relations. And the open-mindedness, broadness, creativity, erudition and generosity that marked his personality furthered a general holistic attitude and an integrative mind. If I am correct about the role individual characters play in the formation phase of new scientific theories, then this suggests that personal virtues and epistemic virtues form a continuum and are indeed hard to separate from each other.

We should not forget that Jakobson's personality also fitted the 'Zeistgeist'. Jakobson minted the term structuralism to capture a wholesale change in science towards anti-positivism and relationalism. He saw parallels between linguistics and other fields of study, such as physics and musicology. Next to this, many attempts to deliver an all-encompassing synthesis could be witnessed in the first decades of the twentieth century. Structuralism appeared to present a solution to the long-standing problem of how to anchor modern science in general culture as it promised to further the causes of solidarity and liberty.

In the case of Jakobson it is not just that the man and his work are one but also that the man, his work and his time are one. While we can draw connections between self, science and society by noting similarities in virtue pursuit, this is not to say that Jakobson's 'project' formed a coherent whole, as tension between integration and separation could be witnessed. It would also be premature to draw the conclusion that relationalism, holism, anti-positivism, a conquering mentality and a preference for epistemic virtues such as integration, generalization and simplification form an inevitable bond. Both holism and structuralism have turned out to be extremely porous notions that have been interpreted in different ways and applied in a variety of disciplines, quite possibly serving the pursuit of different sets of epistemic virtues. A comparison between these various applications with respect to the pursuit of virtues falls beyond the scope of this paper but forms an interesting prospect for further research.

## Bibliography

### *Primary Sources*

Berg, Lev S. 1969 [1922]. *Nomogenesis or Evolution Determined by Law.* Trans. J.N. Rostovtsov. Cambridge, MA: MIT Press.

De Saussure, Ferdinand. 1916. *Cours de linguistique générale*. Paris: Payot.

Huizinga, Johan. 1935. *In de schaduwen van morgen: een diagnose van het geestelijk lijden van onzen tijd.* Haarlem: Tjeenk Willink.

Jakobson, Roman O. 1929. 'Romantické všeslovanství-nová slavistika [Romantic pan-Slavism: a new Slavistics']. *Čin* 1–1: 10–12.

———. 1931. Über die phonologischen Sprachbünde. *Cercle* 4: 234–240.

———. 1941. *Kindersprache, Aphasie und allgemeine Lautgesetze*. Uppsala: Almqvist & Wiksell.

———. 1962. Retrospect. In *Selected Writings I: Phonological Studies*, ed. R. Jakobson, 629–658. The Hague: Mouton.

———. 1971a. The Kazan School of Polish Linguistics and its Place in the International Development of Phonology. In *Selected Writings*, ed. R. Jakobson, vol. 2, 394–428. The Hague: Mouton.

———. 1971b. Musikwissenschaft und Linguistik. In *Selected Writings*, ed. R. Jakobson, vol. 2, 551–553. The Hague: Mouton.

———. 1979. Новейшая русская поэзия: Подступы к Хлебникову [Modern Russian Poetry. Sketch 1: Approaching Khlebnikov]. In *Selected Writings*, ed. S. Rudy and M. Taylor, vol. 5, 299–354. The Hague: Mouton.

———. 1981a. Linguistics and Poetics. In *Selected Writings*, ed. S. Rudy and R. Jakobson, vol. 3, 18–51. The Hague: Mouton.
———. 1981b. Poetry of Grammar and Grammar of Poetry. In *Selected Writings*, ed. S. Rudy and R. Jakobson, vol. 3, 87–97. The Hague: Mouton.
———. 1985a. Einstein and the Science of Language. In *Selected Writings*, ed. S. Rudy, vol. 7, 254–264. Berlin: Mouton.
———. 1985b. My Favorite Topics. In *Selected Writings*, vol. 7, ed. S. Rudy, 371–376. Berlin: Mouton.
Jakobson, Roman O., and Morris Halle. 1956. *Fundamentals of Language*. The Hague: Mouton.
Marinetti, Filippo T. 1909. Manifesto del Futurismo. *Gazzetta dell'Emilia* 5–2–1909.
Smuts, Jan C. 1926. *Holism and Evolution*. London: MacMillan & Co.
Spengler, Oswald. 1918. *Der Untergang des Abendlandes*. Vienna/Leipzig: Braumüller.
Stalin, Joseph V. 1950. *Marxism and Problems of Linguistics*. Online at https://marxists.org/reference/archive/stalin/works/1950/jun/20.htm.
Stumpf, Carl. 1926. *Die Sprachlaute: experimentell-phonetische Untersuchungen nebst einem Anhang über Instrumentalklänge*. Berlin: Julius Springer.

## *Secondary Sources*

Armstrong, Daniel, and Cornelis H. van Schooneveld. 1977. *Roman Jakobson: Echoes of His Scholarship*. Lisse: Peter de Ridder Press.
Baneke, David. 2008. *Synthetisch denken: natuurwetenschappers over hun rol in een moderne maatschappij 1900–1940*. Hilversum: Verloren.
Birus, Hendrik, Sebastian Donat, and Burkhard Meyer-Sickendiek, eds. 2003. *Roman Jakobsons Gedichtanalysen: Eine Herausforderung an die Philologen*. Göttingen: Wallstein Verlag.
Bradford, Richard. 1994. *Roman Jakobson: Life, Language, Art*. London: Routledge.
Chvany Catherine, V. 1984. From Jakobson's Cube as Objet d'Art to a New Model of the Grammatical Sign. *International Journal of Slavic Linguistics and Poetics* 29: 43–70.
Cooke, Raymond. 1987. *Velimir Khlebnikov: A Critical Study*. Cambridge: Cambridge University Press.
Creyghton, Camille M.H.G., Pieter A. Huistra, Sarah Keymeulen, and Herman Paul. 2016. Virtue Language in Historical Scholarship: The Cases of Georg Waitz, Gabriel Monod and Henri Pirenne. *History of European Ideas* 42: 924–936.
Daston, Lorraine, and Peter Galison. 2007. *Objectivity*. New York: Zone Books.
Daston, Lorraine. 2008. Objectivity and Its Critics. *Victorian Studies* 50: 666–677.
DePaul, Michael R., and Linda Trinkaus Zagzebski, eds. 2003. *Intellectual Virtue: Perspectives from Ethics and Epistemology*. Oxford: Clarendon Press.
Flack, Patrick. 2016. Roman Jakobson and the Transition of German Thought to the Structuralist Paradigm. *Acta Structuralica* 1: 1–15.
Friedman, Victor A. 2014. 'Languages are Wealth': The Sprachbund as Linguistic Capital. *Proceedings of the 38th Annual Meeting of the Berkeley Linguistics Society*: 163–178.
Galan, Frantisek W. 1985. *Historic Structures: The Prague School Project, 1928–1946*. Austin: University of Texas Press.
Gardner, Howard. 1985. *The Mind's New Science: A History of the Cognitive Revolution*. New York: Basic Books.
Gasparov, Boris. 2014. Futurism and Phonology: The Futurist Roots of Jakobson's Approach to Language. *Ulbandus Review* 16: 84–112.
———. 2016. Between Methodological Strictness and Moral Appeal: Questions of Language and Cultural Theory in Russia. *History of Humanities* 1: 303–326.

Givon, Thomas. 2016. Beyond Structuralism: Exorcizing Saussure's Ghost. *Studies in Language* 40: 681–704.
Halle, Morris. 1983a. On the Origins of the Distinctive Features. *International Journal of Slavic Linguistics and Poetics* 27: Supplement 'Roman Jakobson: What He Taught Us': 77–86.
———, ed. 1983b. *A Tribute to Roman Jakobson 1896–1982*. Berlin: Mouton.
Holenstein, Elmar. 1975. *Roman Jakobsons Phänomenologischer Strukturalismus*. Suhrkamp: Frankfurt am Main.
Hunt, Horace G. 1983. Slavic Historical Linguistics. *International Journal of Slavic Linguistics and Poetics 27: Supplement 'Roman Jakobson:What He Taught Us'*: 52–56.
Ivanov, Vjaceslav V. 1983. Roman Jakobson: The Future. In: A Tribute to Roman Jakobson 1896–1982, ed. M. Halle, 47–57. Berlin: Mouton.
Karstens, Bart. 2012. Bopp the Builder: Discipline Formation as Hybridization: The Case of Comparative Linguistics. In: *The Making of the Humanities*, vol. 2, ed. R. Bod, J. Maat and T. Weststeijn, 103–127. Amsterdam: Amsterdam University Press.
———. 2017. The Prague Linguistic Circle and the Analogy Between Musicology and Linguistics. *History and Philosophy of the Language Sciences*. Online at https://hiphilangsci.net/2017/03/07/prague-musicology-linguistics/.
Kiparsky, Paul. 1983. Roman Jakobson and the Grammar of Poetry. *International Journal of Slavic Linguistics and Poetics* 27: Supplement 'Roman Jakobson: What He Taught Us': 20–29.
Koerner, Konrad. 1982. The Schleicherian Paradigm in Linguistics. *General Linguistics* 22: 1–39.
Kohrt, Manfred. 1984. *Phonetik, Phonologie und die 'Relativität der Verhältnisse': zur Stellung Jost Wintelers in der Geschichte der Wissenschaft*. Stuttgart: Franz Steiner.
Kuhn, Thomas S. 1962. *The Structure of Scientific Revolutions*. Chicago: University of Chicago Press.
Kursell, Julia. 2010. First Person Plural: Roman Jakobson's Grammatical Fictions. *Studies in East European Thought* 62: 217–236.
Laudan, Larry. 1977. *Progress and Its Problems: Toward a Theory of Scientific Growth*. Berkeley: University of California Press.
Lazard, Gilbert. 2012. The Case for Pure Linguistics. *Studies in Language* 36: 241–259.
Marácz, Laszlo. 2015. *Towards Eurasian Linguistic Isoglosses: The Case of Turkic and Hungarian*. Astana: International Turkic Academy.
McAllister, James W. 1996. *Beauty and Revolution in Science*. Ithaca: Cornell University Press.
———. 2011. What Do Patterns in Empirical Data Tell Us About the Structure of the World? *Synthese* 182: 73–87.
Morpurgo Davies, Anna. 1998. *History of Linguistics*, vol. 4. London: Longman.
Paul, Herman. 2014. What Is a Scholarly Persona? Ten Theses on Virtues, Skills, and Desires. *History and Theory* 53: 348–371.
———. 2017. *Weber, Wöhler, and Waitz: Virtue Language in Late Nineteenth-Century Physics, Chemistry, and History*. In this volume.
Percival, Keith W. 2011. Roman Jakobson and the Birth of Linguistic Structuralism. *Sign Systems Studies* 1: 236–262.
Rudy, Stephen. 1978. Jakobsonian Poetics of the Moscow and Prague Periods. PhD thesis, Yale University.
Sangster, Rodney B. 1982. *Roman Jakobson and Beyond: Language as a System of Signs: The Quest for the Ultimate Invariants in Language*. Berlin: Mouton.
Sériot, Patrick. 2014. Structure and the Whole. *East, West and Non-Darwinian Biology in the Origins of Structural Linguistics*. Trans. A. Jacobs-Colas. Boston: Mouton.
Shapin, Steven. 2008. *The Scientific Life: A Moral History of a Late Modern Vocation*. Chicago: Chicago University Press.
Stankiewicz, Edward. 1983. Roman Jakobson: Teacher and Scholar. *International Journal of Slavic Linguistics and Poetics* 27: Supplement 'Roman Jakobson: What He Taught Us': 17–26.
Steiner, Peter. 1984. *Russian Formalism: A Metapoetics*. Ithaca, NY: Cornell University Press.

Tai, Chaokang, and Jeroen van Dongen. 2016. Anton Pannekoek's Epistemic Virtues in Astronomy and Socialism: Personae and the Practice of Science. *Low Countries Historical Review* 131 (4): 55–70.

Toman, J., ed. 1994. *Letters and Other Materials from the Moscow and Prague Linguistic Circles 1912–1945*. Ann Arbor: Michigan Slavic Publications.

Van Dongen, Jeroen. 2010. *Einstein's Unification*. Cambridge: Cambridge University Press.

———. 2017. *The Epistemic Virtues of the Virtuous Theorist: On Albert Einstein and His Autobiography*. In this volume.

Van Fraassen, Bas. 1980. *The Scientific Image*. Oxford: Clarendon Press.

Van Spaendonck, Jan. 1977. *Belle époque en anti-kunst: de geschiedenis van een opstand tegen de burgerlijke cultuur*. Meppel: Boom.

Waugh, Linda S. 1976. *Roman Jakobson's Science of Language*. Lisse: Peter de Ridder Press.

Winner, Thomas G. 1977. Roman Jakobson and Avantgarde Art. In: *Roman Jakobson: Echoes of His Scholarship*, ed. D. Armstrong and Ch. van Schooneveld, 503–514. Lisse: Peter de Ridder Press.

———. 1984. Review Article Roman Jakobson, The Poetry of Grammar and the Grammar of Poetry. Selected Writings, vol. 3. *International Journal of Slavic Linguistics and Poetics* 30: 159–166.

Worth, Dean S., Edward Stankiewicz, Van Schooneveld, H. Cornelis, and Walter Vickery. 1983. † Roman Jakobson 1896–1982. *International Journal of Slavic Linguistics and Poetics* 27: 7–10.

# Johan Rudolph Thorbecke's Revenge: Objectivity and the Rise of the Dutch Nation State

**Ad Maas**

**Abstract** In this paper the emergence of objectivity as a leading epistemic virtue in the Dutch academic sciences is discussed in the context of the social-political developments that took place in Dutch society in the nineteenth century. I argue that the emergence of objectivity was closely connected to the rise of a formally-ruled and integrated nation state. My analysis is based on a case study, which involves three historians: Philip Willem van Heusde, Johan Rudolph Thorbecke, and Robert Fruin. For each historian, I show how their epistemic virtues (more particularly, those involved in the rise of objectivity) reflected, or contradicted, the Dutch social-political organization.

**Keywords** Epistemic virtues • Objectivity • Philips Willem van Heusde • Johan Rudolph Thorbecke • Robert Fruin • History of historiography • Dutch history

## Introduction

Studies on epistemic virtues in the nineteenth century generally do not seek to relate these to the broader societal contexts in which scholars conducted their research and disseminated their views.[1] In this paper, in contrast, I aim to demonstrate that epistemic virtues, not to say scholarly practices in general, can be understood only

[1] Much of this burgeoning scholarship on epistemic virtues is 'internalist' to the extent that it pays little attention to societal contexts in which scholars conducted their research and disseminated their views. Recent literature on one of this paper's main protagonists, Robert Fruin, is a case in point. Fruin's epistemic ideals have been examined in some detail (Tollebeek 1990). The virtues he practiced and preached have been compared to those of other Dutch humanities scholars (Paul 2011, 2012; Tollebeek 2011). Also, there has been constant attention to the way Fruin's epistemic virtues translated into working habits and educational practices (Tollebeek 2015). However, neither these studies, nor the long-standing historiography on Fruin's political views (Smit 1958; Vermeulen 1959), raised the question how Fruin's epistemic virtues reflected or presupposed a legal-political order and/or a social context of the sort characteristic of late nineteenth-century Netherlands.

A. Maas (✉)
Museum Boerhaave, Leiden, The Netherlands
e-mail: admaas@museumboerhaave.nl

 
J. van Dongen, H. Paul (eds.), *Epistemic Virtues in the Sciences and the Humanities*, Boston Studies in the Philosophy and History of Science 321,
https://doi.org/10.1007/978-3-319-48893-6_11

against the background of social-political orders that favored some virtues over others. More particularly I will contend that the emergence of objectivity as a leading epistemic virtue in the Dutch academic sciences in the nineteenth century (in the broad sense of *Wissenschaft*, the humanities and social sciences included) was connected to the rise of a centralized and formally-ruled Dutch nation-state.

My case studies involve three nineteenth-century Dutch historians. The first is Philip Willem van Heusde, a leading academic during the reign of King William I (1813–1840). The second is Robert Fruin, the influential professor of national history of the last decades of the nineteenth century, who has been regarded by later generations of historians as the 'father' of the historical discipline in the Netherlands.

The epistemic virtues Fruin endorsed and his professional convictions in a more general sense, made a marked contrast with those of Van Heusde. The latter searched for truth using his erudition, personal experience and hard-won intellectual skills. Fruin, in contrast, was more skeptical of intervention on the scholar's part. Like some of his contemporaries, he tended to conceive of such interventions as an obstacle, rather than as a tool for arriving at the truth. Ideally, objects of research (nature, society, history) should be registered without the intervention of judgments and simplifications on the part of the scholar. This is to say that Fruin, unlike Van Heusde, was committed first and foremost to objectivity. I will argue that both figureheads represent, at least to some degree, the Dutch scholarly and academic culture of their times.

In broad lines, this article follows the pattern set out by Lorraine Daston and Peter Galison in their book on the emergence of objectivity as a leading epistemic virtue in observational sciences, and Daston's subsequent work on objectivity in the humanities. Yet, while Daston and Galison deem wider societal developments 'inadequate' explanations for 'objectivity' and 'only remotely relevant', I will try to demonstrate their interconnectedness (Daston and Galison 2007, quotations on 35–36; Daston 2014). This is not a far-fetched hypothesis: as the nation's intellectual and moral authorities and as teachers of the nation's ruling classes, professors were part of (or close to) the top ranks of the sociopolitical hierarchy.[2] Several examples in this paper show that their discursive style and professional virtues resembled those of the leaders of the nation and were influenced by changes in the socio-political sphere.

This is particularly clear in the case of our third historian, Thorbecke, who is the pivotal figure of this article. He is mainly known as the statesman who – as the architect of the liberal constitution of 1848 – almost single-handedly laid the foundation for the modern Dutch polity. However, before becoming the leading politician of the Netherlands, Thorbecke was active as a scholar, working in philosophy, law and history. His troublesome relationship with the Dutch scholarly community provides interesting insights into the relation between the socio-political context and the rise of objectivity. At a time when the revered scholar Van Heusde made an

[2] In the beginning of the century professors still ranked below the patriciate socially, at the end they had risen to the top of the social pyramid (Caljé 2009, 196–275). On political leaders and university training see also Caljé 2009.

indelible impression on his contemporaries, Thorbecke pursued a different kind of scholarship, one inspired by German examples, which proved to be incompatible with the values of Dutch culture. To some extent Thorbecke can be considered a forerunner of an 'objective' historian.

Ironically, Thorbecke's scholarship would have been much more appreciated in the scholarly climate which his political reforms eventually helped to establish, the climate also in which Fruin flourished. The implementation of Thorbecke's liberal constitution initiated a profound societal transformation, in the course of which Dutch society changed from a largely locally organized, oligarchic class society to a socially more open, formally-ruled nation-state. In this article, which is mainly based on the study of secondary sources, I will argue that the emergence of objectivity in this period was related to this process. Exploring the cases of Van Heusde, Thorbecke, Fruin and the epistemic virtues they espoused will furthermore help us to reveal in a broader sense some particularities of Dutch scholarly practice and convention, and their compatibility with the organization of Dutch society.

## Adequately Deploying Reason

My first case is that of Philip Willem van Heusde (1778–1839). This Utrecht professor, who was called the 'Praeceptor Hollandiae', made an enormous impression on his contemporaries. As an influential scholar he can be considered a figurehead of the Dutch academic community in the first half of the nineteenth century. Van Heusde was a classicist, philosopher and historian. He was an adherent of Plato, and of common-sense philosophy, as well as being a pious, though not dogmatic Christian. According to Van Heusde, it was thanks to God-given reason that progress was seen to be made in human history. This was the main – and optimistic – leitmotiv of his historical perspective. (Lakke 1908; Sassen 1959; Roelevink 1986, esp. 291–293, 311–315; De Valk 1989, 13–35; Van Raak 1999) Van Heusde's main work was *De Socratische school* (The Socratic School) (4 vols., 1834–1839). In this encyclopedic work he provides an overview of how arts, sciences and religion have enriched human insight. In an elaboration on 'music, poetry and visual arts', for instance, he analyses – largely by drawing on examples from ancient Greece – how these have contributed to the ennoblement of human kind. Van Heusde generally supported his views without bothering too much about refinements and details (De Valk 1989, 22). Among those who came under the spell of Van Heusde were the theologians who founded the *Groninger Richting*, the theological school which has been considered the main (if not the only) innovative intellectual Dutch movement of the first half of the nineteenth century. Inspired by Schleiermacher's theology and Van Heusde's belief in progress, the Groningen theologians considered the person and conduct of Jesus as a model for achieving ethical and intellectual perfection, focusing less on his sacrificial death for the sins of humankind and on ecclesiastical doctrines (Vree 1984).

Van Heusde fitted historical facts into his views on the progress and destination of humankind, relying on his authority as a scholar and making few attempts to let facts speak for themselves. Van Heusde selected and idealized his data in order to distill their essence, the essence as *he* saw it, to be sure. He was certainly not the type of scholar who conceived of his own intellectual intervention as an obstacle in the way of finding scholarly truth. According to his epistemology, empirical facts yielded the scholar but a faint resemblance of the truth, which can only be seen by the intervention of the mind, or – in Van Heusde's expression – the 'eye of the soul' (*zieloog*) (Lakke 1908, 166–169).

Scholarly truth thus bore the mark of the scholar's personality. Van Heusde, who was also one of the main spokesmen for the pedagogical values of university training, reflected this in his educational ideals, which encouraged 'the minds of the listeners to judge, to separate and combine, to search regularly and to find the truth. This, who will deny, is the true spirit of academic education' (cited in Bagge 1998, 69).

The skill of making good judgments was not easily won. In *De Socratische school* Van Heusde extensively writes about his own intellectual and spiritual development. 'About the mutual relation and the objective of all arts and sciences I have thought, from my earliest youth on, a lot and often', he writes in the opening sentence of his opus magnum, and what follows is a pages-long description of his own intellectual development. It apparently took decades of intellectual effort and experience to underpin his work with scientific legitimacy and authority: '(...) it is now almost thirty years ago, that I came to experience this feeling, hope and expectation (…) and still, friends, though I have already entered my thirty-fifth year, I still doubt, whether I have come to sufficient ripeness (…)' (Van Heusde 1860). Van Heusde largely seems to resemble Daston and Galison's sage 'truth-to-nature' scholar 'whose well-stocked memory synthesizes a lifetime of experience' (Daston and Galison 2007, 44).

To fully understand Van Heusde's use of the scholarly mind, it is important to emphasize that in his epistemology, finding truth was not an act of reason alone. Reason needed the guidance of moral values: what was true was also morally right. In this respect Van Heusde was a critic of Enlightenment thought, and of Immanuel Kant (1724–1804) in particular. Whereas the latter claimed that all humans possessed inborn faculties of cognition, Van Heusde contended that the state of these faculties depended on the degree of moral completion of the individual, which would only be achieved when one was capable of harmonizing reason, feeling and will. Only then 'intuitive, immediate and evident knowledge', would be obtained (not only applicable, incidentally, in scholarly matters, but also in moral issues and for political decision-making) (Van Raak 1999, citation on 32). Van Heusde certainly did not stand alone in this respect. Humanities professor Johannes Pieter van Capelle also contended that historiography was pointless without a moral component. The historian should judge the past with affection and modesty. 'Thus he disseminates the seeds of the truth, beautiful and good, from generation to generation' (Van Berkel 2008, 162).

The linking pin between morality and reason was the use of common sense, a key term in Van Heusde's philosophy as well as in the wider intellectual culture in the

Netherlands at the time (Sassen 1959; De Quay 2000, 106–109). The philosophy of common sense (inspired by David Hume and Frans Hemsterhuis) cemented the elements of Van Heusde's philosophy into its rather eclectic whole. Common sense could be instrumental as a powerful epistemic tool, it 'always strove to take the middle course, by constantly resolving, accommodating and cautioning whenever there was only the slightest risk of straying off course' (Lakke 1908, 44). Common sense thus led the scholar to form morally and reasonably 'true' judgments.[3]

The capability of making sound judgement, in conclusion, was reserved for those who had been through a profound moral and intellectual development, and knew how to apply common sense. Such sage persons were 'intuitively' able to produce correct knowledge and statements. As Jo Wachelder claims, scholarship in the Netherlands in the first half of the nineteenth century was 'the cumulative development of the person', a 'life-ideal' (...) 'bound to the person' (…) implying 'civilization, patriotism, good citizenship, morality and knowledge' (Wachelder 1992, 216).

## Paternalism and Moralism

Van Heusde's scholarship and its moral dimensions were congruent with the Dutch political organization during the reign of William I. The Dutch polity was strongly focused on the person of the king, who deployed an autocratic and paternalistic style of governance. Rather than by law, he ruled by (royal) decree (Tamse and Witte eds. 1992; Van Sas 2005, 413–435; Lok 2014).

William I was often metaphorically referred to as the 'father' of the country (*landsvader*), echoing the principles of 'oeconomy'. In the oeconomic ideal of governance, society was considered as a kind of household. At the top was a 'fatherly' figure who exercised a stern but righteous rule over his subjects. He was supposed to know what was best for them – based on his own judgment – and had to set an example with his own virtuous behavior (De Valk 1992, 77–79; Van Sas 2005, 464–468; on oeconomy: Van Driel 2016). The ruler of the post-Napoleonic Restoration period was a 'moral administrator', whose reputation had to be morally impeccable (Lok 2014, 62–63). The conviction that fair rule was based on moral arguments, tradition and intuition remained widespread until at least 1850 (Houwaart 1991; Klep, Maarseveen and Stamhuis eds. 2008; there esp. Klep 2008).

Persons were often more important than the positions they held and the administration was in the hands of a small circle of royal confidants: formal and informal power structures were intertwined and administrators – often accumulating several

[3] It is interesting to note that the Dutch academic community's adherence to commonsensical reasoning was related to the circumstance that Dutch culture was obsessed with the idea of 'keeping the middle' following the chaotic revolutionary period. A 'cult of simplicity' even took hold of Dutch culture, in which alleged 'national' values such as honesty, modesty, diligence, solidity, piety, moderation, perseverance and – indeed – common sense were cultivated. Several historians have qualified this Dutch culture as a 'Biedermeier culture' (see Kossmann 1995, 56; Johannes 1997, 79–86; Aerts 1997, 40; Cossee 1988, esp. 208–225; Maas 2010).

positions of authority –in certain respects served the king rather than the state as an ideal-typical Weberian bureaucrat. What is said about Van Heusde's work can also be said about William's kingdom: 'The result was an unstable whole, held together only by (...)[his] personality' (De Valk 1989, 31).

The scholarly practice that was grounded in the authority of the individual scholar, in short, found an equivalent in the oeconomic administrator. On the strength of experience and personality, the scholar knew what was true, while the administrator knew what was best. The moral compass that guided Van Heusde and his colleagues was mirrored by a polity that was often based on moral arguments rather than led by formal rules and legislation. In the 1850s Van Heusde's views would actually be used by conservatives who were led by the hotheaded Utrecht chemist Gerrit Jan Mulder (1802–1880). The latter aimed to restore the personal authority of the king at the cost of the constitutional system that had been introduced after the liberal reforms of 1848 (Van Raak 1999).

## Local and Verbal

So far we discussed the epistemology of Van Heusde's historical and philosophical views. A different aspect is how these views were communicated. In this respect it is interesting to note that especially Van Heusde's academic teaching made a lasting impression on his followers. 'He made an impression with his education, his infectious enthusiasm, and the inspiring effect of his personality. Only in the second place were his publications influential', as one biographer states (De Valk 1989, 13). The same goes, by the way, for the members of the *Groninger Richting* mentioned earlier, who more effectively pursued their views in their sermons in Groningen's St. Martin's Church than in scholarly, written publications.

Tellingly, even Van Heusde's scholarly, written texts were composed as if he was speaking: 'Young men, you who apply yourselves to practicing the arts and sciences, tell me, how did you fare so far solving these issues? For it is important to know where the road leads to if one intends to enter upon it' (Van Heusde 1860, 2). It is as if he is preaching to an audience, rather than transmitting scholarly information.[4] Van Heusde expressed his own personality when communicating his views.

This personal and direct style of communication accorded with the Dutch cultural landscape, in which public debates and scholarly interaction often took place in the semi-closed arenas of the numerous learned societies that profoundly shaped social, cultural and scholarly life from the second half of the eighteenth century onwards. Here, scholars rubbed shoulders with educated and affluent citizens (see, for instance, Kloek and Mijnhardt 2001, 103–126). Dutch professors – who as members of the Republic of Letters of course also fostered international contacts – were remarkably active in the local societies, not only as members, but also as

[4] It has been noted that Dutch poetry in this period was also written as if it had to be recited, rather than read from paper (see Van den Berg 1984).

driving forces, leaders and founders of quite a few of them. We know that Van Heusde was involved in the foundation of a *leesbibliotheek* (reading room) in Utrecht. He was also the director of the *Provinciaal Utrechtsch Genootschap* (Provincial Utrecht Society), though in his case societies do not seem to have been an important forum to disseminate his ideas (Anon. 1840).

Dutch cultural life had a strong local orientation. The cultural fragmentation resulted from the circumstance that the Netherlands, in spite of its formal unification, remained in practice 'an archipelago of areas and societies', with a comparatively low degree of integration between the several regions and towns (compared for instance to the end of the century) (Knippenberg and De Pater 1990; De Rooy 2002, 15–45). One cause of this was the deficient infrastructure, another one had to do with the system of local taxation, which frustrated economic integration. However, it also reflects William I's only partial success in bringing the country under central rule. He lacked the means to exercise his power at a local level and was therefore still forced to rely on traditional local power structures (patricians, aristocrats, ecclesiastic institutions). Centuries of particularism were not simply brought to an end with one stroke of the pen (Caljé 2009).

The learned societies reflected the prevailing ideal of sociability. According to this ideal, it was only in a collaborative effort, by joining forces, that personal development as well as scholarly activities could be furthered. The scholarly practice was seen as a collective practice of debating, undertaking experiments and engaging in field expeditions together. In contrast, the scholar who worked in isolation in his study was outright considered to be a 'misanthrope'. 'A professor should not write much', the Groningen professor (and Van Heusde's student) Cornelis Star Numan (1807–1857) claimed, 'his time is reserved for students' (Huizinga 1951, 135).

This ideal of sociability corresponded with a scholarly culture of direct, often verbal communication. Indeed, Dutch scholarly culture was largely an oral culture. As we have seen, this is clearly reflected in Van Heusde's argumentation and rhetorical style. The oral orientation of the scholarly practice, which is also indicated by the importance attached to eloquence (*welsprekendheid*) in the academic curriculum, is something which has largely been overseen in historiography, but must have made a mark, not only on the style and rhetoric of the scholars, but also on their type of argumentation and techniques of persuasion.[5] Obviously, in direct, verbal communication personal authority and persuasiveness play a stronger role than in the formal, written reporting of scholarly results. Like the Dutch polity, the fragmentation of Dutch culture favored a personalized style and method.

So far, we have seen that the social and political organization under King William I was compatible with the scholar whose aspirations to truth strongly depended on his personal authority. Conversely, as our next case will show, this societal structure turned out to be adverse to a more detached kind of scholarship.

[5] Van Berkel 2008 (118–119 and 155–159) briefly mentions the oral orientation of Dutch scholarly culture in the first half of the nineteenth century; on oral culture in poetry, see Van den Berg 1984).

## *Etwas für sich*

There had been one dissenter among the numerous admirers of Van Heusde, someone who did not appreciate Van Heusde's scholarly work, who disparaged his scholarly style, who even disliked him personally. This someone was a marginal philosopher who, however, one day would almost singlehandedly lay the foundation of the modern Dutch political system: Johan Rudolph Thorbecke (1798–1872). For Thorbecke, Van Heusde embodied the oppressive companionability of the societies, the perfunctory nature of scholarly expositions, the lack of profound scholarly debate that characterized Dutch scholarly culture and even Dutch culture at large. Thorbecke, to be sure, hated this culture (Drentje 2004).

Although the Thorbecke merchant family had already lived in the provincial city of Zwolle for generations, it never shed its German roots. Johan Rudolph did not manage it either. Already as a student – and certainly following a two-years' stay in Germany – he came under the spell of German idealistic philosophy, *Altertumswissenschaft* (science of Antiquity) and history. It almost cost him his academic career. The Dutch scholarly community, headed by Van Heusde, did not appreciate any philosophizing of an epistemological nature that transcended the reach of common sense and aimed at personal introspection rather than societal benefit or religious purposes. Kantian transcendental philosophy – and anything that went beyond – was simply dismissed as 'mysticism'. There was no room for a philosophy independent of religion.

Thorbecke was a promising student, destined to occupy a chair in Leiden. He was sidelined because of his associations with German idealistic philosophy, however, which took him to Ghent (then part of the United Kingdom of the Netherlands) where he was appointed to an endowed chair, not in philosophy but in European history, international relations and political science. In spite of his reputation, Thorbecke was certainly not uncritical of idealistic philosophers, which he often considered too speculative, and he did not want history to be straitjacketed in philosophical systems, such as Hegel's or Krause's. He found inspiration in the empirical approach of Göttingen, especially in the work of Arnold Hermann Ludwig Heeren (1760–1842). Unlike Van Heusde, and decades before Fruin's 'impartiality', Thorbecke resolved to regard every historical era according to its own standards and merits: it was not the task of the historian to condemn or approve (Drentje 2004, 27–266).

Thorbecke would never make a monumental contribution to scholarship, not even when he finally did become professor in Leiden in 1831. His true destiny lay elsewhere. However, in an intriguing analysis of the scattered publications in which Thorbecke unfolded his historiographical views, the Dutch historian Ernst Heinrich Kossmann points at the far-reaching similarities between Thorbecke and Leopold von Ranke (1795–1886) – who was Thorbecke's senior by three years (Kossmann 1987; see also Drentje 2004, 213–266). These similarities concerned not only their view on historiography, but also the philosophical and religious foundations that buttressed these views, their sources of inspiration (in particular Johann Gottfried

Eichhorn (1752–1827)), and even their careers, personal lives and their political views (which only began to diverge after 1840). Thorbecke, like Ranke, regarded history from an organic-idealistic point of view. Consequently he considered historical phenomena as part of a larger whole. However, these phenomena (cities, persons, eras) also had their own particular significance, or 'individuality'; they had 'etwas für sich' (something of its own) as Thorbecke formulated it. The significance of history largely rested in these phenomena, which could be uncovered by rigorous empirical research, through the critical study of sources (it was here that the Göttingen empirical approach made itself felt). Only then could the coherence of a historical period become visible. History, in contrast, could not be understood by making rash generalizations, and Thorbecke (again like Ranke) therefore dismissed both idealistic systems such as that of Georg Wilhelm Friedrich Hegel (1770–1831), and universalistic, enlightened views on the historical development of mankind such as that of Van Heusde.

In contrast to Van Heusde, Thorbecke was no 'teacher of life', he nowhere emphasized history as a means to further patriotism or moral values (Drentje 2004, 235–237). Both would be incompatible with the epistemic restraint he imposed on it, based on regarding historical events and phenomena according to their own merits and standards. Needless to say, common sense was not regarded by him as a tool to arrive at philosophical and factual insights. Following Daston's claim – which will be discussed below– that Ranke may be considered a founding father of historical objectivity thanks to his critical use of historical sources, Thorbecke, too, may be regarded as a forerunner (though, of course, he was not as influential as Ranke). 'Maybe he could have become a Dutch Guizot or Ranke', Drentje speculates, 'instead, the times would have to wait for Fruin' (Drentje 2004, 234). Indeed, Thorbecke's approach to historical research clearly resembled Fruin's historiography more than Van Heusde's. As will also become clear in the section 'History from a distance', the main difference with Fruin would be that Thorbecke's historiography, despite the epistemic restraint embodied by the latter's 'etwas für sich', was idiosyncratically embedded in his metaphysical views.

One of the main sources for Thorbecke's historiographical views is the proposal for a general history of the Netherlands which he submitted to a competition that was issued by Royal Decree in 1826. Unlike the 43 other participants, Thorbecke did not present an outline of Dutch history, but instead proposed to initiate a series of source publications. As they were being prepared, sub-studies could be conducted as a prelude for an envisaged general national history. Thorbecke's grassroots approach did not fall on fertile ground. Much to the surprise of later historians, his proposal was not even selected among the five best. As Jan Drentje has found out, the most influential member of the jury probably was actually Van Heusde. Incidentally, in 1902 a series of source publications, the *Rijks Geschiedkundige Publicatiën,* was indeed started. It explicitly elaborated on Thorbecke's blueprint proposed some 75 years earlier (Kossmann 1987, 19–20; Drentje 2004, 231–238).

As becomes clear in Drentje's intellectual biography of Thorbecke, the irreconcilability between Thorbecke and the Dutch scholarly community involved more than just philosophical arguments. It also concerned a more fundamental incompatibility between a principally individualistic philosopher and a society-culture, informed by moderate Christian and enlightened ideas. As has been noted, Thorbecke's predilection for the empirical study of sources was related to his metaphysical view on the individuality of every historical entity. All historical phenomena were part of a larger, organic whole, but at the same time were also 'etwas für sich'. According to Thorbecke, however, the 'etwas für sich' similarly also applied to the individuality of a person in relation to his or her social environment. In his view, genius and originality relied on a personal, autonomous 'sphere' which, however, was in danger of being intruded on by others. Drentje suggests that his willingness to be 'etwas für sich' underpinned Thorbecke's personal drive for self-realization and the fulfilment of his ambitions (Drentje 2004, 120–123 and 326).

Even though Thorbecke's philosophy may be somewhat arcane to modern observers, the connection it makes between epistemic detachment and detachment in the personal domain, indicates an intriguing relation between the rise of 'objectivity' and the emergence of a more individualistic culture. Thorbecke's individualistic drive for self-realization, indeed, was difficult to reconcile with the societal configuration of Van Heusde's Netherlands, in which social intercourse was strongly regulated by outward appearances: 'In a small scale class society, where much depended on personal relations, prestige, authority and family name formed an indispensable commodity.' In this context the individual naturally largely coincided with his social role (Aerts 1998; Aerts 2002, quotation on 23; Drentje 2004, 121–123; Maas 2010). Little room was left for being 'etwas für sich'. For years, the disengaged Thorbecke – mentally residing in a nowhere land between Germany and the Netherlands – lived the life of an ascetic, misunderstood scholar.

The gap that separated Thorbecke from Van Heusde, was also the gap that separated the scholarly cultures of the Netherlands and Germany, where the rulers had released a small group of civil servants from their traditional bonds to society to serve their expanding bureaucracies. This group, non-existent in the Netherlands, developed a particular individualistic and meritocratic lifestyle and became known as the *Bildungsbürgertum*. It was this attitude, as I have argued elsewhere, that from about 1800 onwards encouraged German scientists to undertake an innovative approach to the sciences and guided their increasing focus on research, distinguishing them in this respect from their Dutch counterparts. It is remarkable how many of the Dutch scientific pioneers of the last decades of the nineteenth century drew on German examples (Maas 2010, 91–95).

The opposition Thorbecke met concerned his overall scholarly attitude, style and preferences. Although a form of 'objectivity' was inherently connected to his scholarship, the resistance against him provides no explicit evidence that the wider societal context and scholarly objectivity *as such* were completely incompatible. Drawing on Thorbecke's and Van Heusde's cases, however, it seems safe to conclude that the social (and even mental) conditions of Restoration–Netherlands best suited a scholarship based on personal authority. Yet, profound societal changes were to come with the advent of 'objectivity' as a leading epistemic virtue in the Netherlands. The remarkable thing is that these would be initiated largely by Thorbecke himself.

## Objectivity and Society

From the moment Thorbecke entered Dutch parliament he presented a rather indefinable figure to his contemporaries. Rather than relying on personal authority and social standing, he underpinned his views by means of his scholarly ideas. Based on his 'organic-idealistic' philosophy, he observed a gradual, liberal evolution taking place in the Netherlands, driven by the political emancipation of its citizens. This revealed itself by a growing demand to have a say in public matters and an increasingly powerful public opinion. Thorbecke's organic liberalism promoted personal freedom, individual initiative, self-regulation and open government on the one hand and restrained state intervention on the other hand. 'Thorbecke', Drentje writes, 'in fact individualized all aspects of the polity'. If every individual, within the legal framework, would contribute according to his own capabilities, the whole of society would grow. A powerful central state was needed to provide and maintain this legal framework. In this way, to express it in a Thorbeckian manner, 'the creative powers' in society would be released (Drentje 2004, 327–449, quotation on 402; Te Velde 1999).

The turbulence which swept over Europe in 1848 did not leave the Netherlands unaffected. Feeling threatened by the dramatic events in the German states and France, the whimsical King William II (who had succeeded his father William I in 1840) eventually gave Thorbecke and his liberal companions the opportunity to draft a new constitution. This constitution, whose main architect was Thorbecke, was complemented by organic laws that changed the Dutch polity profoundly. They curtailed the power of the king and the oligarchies made up of national, provincial and municipal elites. In order to combat provincialism, oligarchy and clientelism, Dutch liberals promoted a strong, central state. The personal style of rule was replaced by a formalistic approach – moral rule was to be replaced by formal rule, and public affairs had to be strictly separated from personal interests. 'Liberalism', in the words of Henk te Velde, 'gave the final impetus to the modern, objectified state apparatus' (Te Velde 1992, 23). Furthermore, a lasting separation of state and church and freedom of press were among the principal measures that would optimize personal freedom and release the 'creative powers' of the Dutch citizen (De Haan and Te Velde 1996; Drentje 2004).

Thorbecke's reforms stimulated national integration by replacing formal and informal local rules and customs with uniform municipal and provincial laws. In addition, local excises were abolished, new waterways and railways improved infrastructure, a telegraph network was established and a new 'Postage Law' (*Postwet*) reduced postal rates by on average 50% (Drentje 2004, 417–419). Together with increasing social mobility, these measures paved the way for the rise of mass-society at the end of the century (Van Holthoon 1977, 100–137; Caljé 2009). It was in this context of the emerging nation-state that the locally organized, largely verbal scholarly culture lost its significance. The development of an 'open national public sphere' had already set in earlier, taking root with the emergence of widely read national periodicals such as *De Gids* (founded in 1837) (Aerts 1997, 83–84).

From then on, academic scholars found themselves definitely working in the framework of a nation state (Maas forthcoming). As we will see in the next section, their epistemological orientation would start to change.

## Objectivity in Governance

In his book *Trust in numbers*, Theodore Porter addresses the communicative abilities of figures, graphs and formulas and other carriers of impersonal, 'objective' knowledge. As their interpretation relies on universal rules, they can move easily across geographic and social boundaries. Quantification, according to Porter, becomes a common language when the boundaries of local society and scientific community are transcended: 'quantification is a technology of distance'. Objectivity in Porter's view serves as an alternative to personal trust (Porter 1996, quotation on ix).[6] As I set out in the previous section, boundaries were transcended in the Netherlands especially after 1850 – not only in a geographical sense, but also in the political domain.

After 1850 the 'statistical mindset' (as Klep, Maarseveen and Stamhuis have dubbed it) started to permeate Dutch governance, advocated by liberals who sought to further openness and transparency of government on principal grounds. Surveys on the working conditions in factories and on the agrarian crisis of the 1880s would lead to the permanent acceptance of statistics as a valid administrative tool, thereby defeating conservative antagonists, who denied its validity and instead advanced moralistic arguments (poor living conditions and social misfortune resulted from morally reprehensible behavior), and arguments based on intuition and tradition as suitable tools to deal with issues of governance. The rise of the statistical mindset also indicated the demise of the personal style of government that had prevailed before Thorbecke's reforms accelerated national integration and increased the scale of governance: small, local and closed circles of administrators were replaced by a form of open government in which – in principle – all Dutch citizens had a say (Klep et al. 2008).[7]

An interesting case in point is that of the hygiene movement, which started to gain momentum in the 1850s. Hygienists looked for new ways of facing the devastating consequences of epidemic diseases. Debates on the underlying causes used to be based on personal and local impressions of individual physicians and badly comparable data. The hygienists strove for a more 'objective' approach. Statistical data, set down in graphs, would reveal patterns that could give insight in the circumstances in which diseases were able to spread, without having to speculate about their causes. Dutch hygienists assembled municipal mortality rates in mortality

[6] It should be noted that Porter's use of objectivity is less strictly defined than Daston and Galison's (see p. ix).

[7] Incidentally, because of their reservations about state intervention, liberals themselves were also slightly ambivalent about statistical surveys.

atlases and related these to population density, welfare levels and geographical characteristics. Their outlook clearly was national, rather than local. From the 1880s their conclusions translated into actual improvements in public hygiene. Interestingly, in this example of the hygienists' use of (in particular) statistics to support both scientific analysis and medical regulation, 'objectivity' in (medical) science and in government coalesced (Houwaart 1991). Indeed, at the time the statistical mindset was hotly debated in Dutch politics, scientific objectivity began to emerge in Dutch science.

## History from a Distance

In the last decades of the nineteenth century, a wave of objectivity washed over the academic sciences in the Netherlands, or – more cautiously stated – at any rate numerous manifestations characteristic of objectivity became mainstream in several academic branches: intellectual and emotional detachment, standardization, automatic procedures for registering, reliance on statistics and other methods of quantification became part and parcel of Dutch scientific practice.[8] Physicians adopted medicine as it was practiced in Germany and France: based less on the physician's personal judgment and more on statistics and instruments. Franciscus Cornelis Donders (1818–1889) is a crucial figure in this development, as were the hygienists referred to in the previous section. In linguistics, the positivist literary historian Jan te Winkel (1847–1927) was a key figure, as was Robert Fruin in history–as we will see (he was actually influenced by Donders as a student). In physics Heike Kamerlingh Onnes (1853–1926), with his famous adage 'door meten tot weten' (from measurement to knowledge) is a paragon of science, being a researcher who relied as little as possible on his imagination. He was preceded (in the sciences) by Frederik Kaiser (1808–1872), who introduced precision measurements in astronomy. Additionally, we could mention formalism (*legisme*) in law, which stressed the prerogative of legislation to limit the interpretation by persons who studied or applied the law (Krop 1992; Theunissen 2000, 99–129; Tollebeek 2011; Van Delft 2005; Zuidervaart 2011).

Like Van Heusde in the post-Napoleonic Restoration Period, Robert Fruin (1823–1899) would become the leading figure of Dutch historiography in the last decades of the century. He was appointed to the first chair in Dutch national history (*vaderlandse geschiedenis*) in Leiden in 1860, and so occupied (in the words of writer Conrad Busken Huet) 'a pulpit in a cathedral with a nation as his audience'. Yet, in contrast to Van Heusde, Fruin did not strive to become the 'nation's schoolmaster', in fact his scholarship was not primarily aimed at an audience at all. Fruin did not produce impassioned speeches that mesmerized readers and listeners, his prose was sober and straightforward. In fact, he did not write much at all.

[8] A more complete picture of the rise of scientific objectivity in the academic sciences in the Netherlands is something that remains to be studied in greater detail.

History was for Fruin first of all a critical evaluation of sources. If the source material was studied profoundly and yielded a clear picture of *wie es gewesen war*, then the principal goal was attained (Aerts 2010, quotation Busken Huet on 195; Tollbeek 2010, 22). Fruin's approach to historical scholarship is reminiscent of Thorbecke's, under whom he had actually studied, and was largely in line with the *Quellenkritik* as it had been developed by German scholars in the nineteenth century (with Fruin's example Leopold von Ranke as their towering figure).

Lorraine Daston identified an essential element of what she considers to be the core of historical objectivity in the method of *Quellenkritik*, namely a fundamental awareness that sources are principally defective and distorted by the historical agents who were involved in their production and usage. Decoding these sources requires methods to establish what they were actually about, and to ascertain how these have been 'colored' by contemporary conceptions and by the author of the source: 'Nothing about a source was self-evident' (Daston 2014, quotation on 32). Thus, the method of *Quellenkritik* strove to achieve 'objective' truth about the past (as permanently elusive as the goal might be). Less tangible, but additionally part of historical objectivity, Daston continues, was a 'vaguer but nonetheless strongly felt value of scientific restraint, which judged precisely how far evidence at hand could be pushed and refrained from pushing it one whit further.' The historian must resist the temptation to generalize prematurely and to 'edify or entertain at the expense of his hard-won facts'. It is not possible here to do full justice to Fruin's scholarship in all its nuances and ambivalences. It is safe to conclude, however, that his focus on source criticism and his restrained style (though in practice perhaps occasionally less strict than prescribed by his own ideals and principles), are largely in agreement with Daston's criteria for historical objectivity.

Seemingly in line with his focus and style was Fruin's defense of 'impartiality'. Already during his student days he noted with disapproval how scholars like Guillaume Groen van Prinsterer (1801–1876) and Willem Bilderdijk (1756–1831) harnessed their colored visions on the national past to settle their own scores on current political and religious issues. It was, according to Fruin, not the task of the historian to side with any historical character, party or event. The historian's task was to evaluate, from a distance, both the merits and the shortcomings of historical agents. This even applied to the pre-eminent bad guys in Dutch history, the Catholics (Tollebeek 1990, 15–67).

A crucial element in Daston's argumentation, however, is that – counterintuitive as it may sound – 'objectivity' cannot be identified with 'impartiality'. Whereas 'objectivity' relates to a scientific method based on the virtue of scientific restraint, 'impartiality' concerns a 'judicial value', aiming to arrive at a sound and fair judgment on a historical event or situation. As Daston shows, nineteenth-century German historians were well aware of this distinction. She mentions the cases of the German historians Heinrich von Sybel (1817–1895) and Georg Gervinus (1805–1871) who regarded history as a means to support the legitimacy of the German empire, and characterized Ranke's impartiality as 'colorless' and a 'sign of political impotence'. At the same time, however, they endorsed and practiced his 'objective' historical-critical method (Daston 2014).

Following this argumentation, Fruin's objective historiography, then, was not inherent to his ideology of impartiality, about which several historians have argued that it was related to his liberal conviction that an open exchange of rational, scientifically grounded points of view would serve to bridge the gap between different ideologies. 'Fruin made from history a kind of parliament in which he allowed every event and every party to speak out with its full weight', as his later critic W. G. C. Byvanck stated. National reconciliation, with the king as a connector, was Fruin's political ideal. His impartiality was all but value-neutral and was colored by his political views (although Fruin himself would have denied this) (Te Velde 2010, 38–59, citation on 46).

Consequently, it was not Fruin's (somewhat problematic) neutrality, but his meticulous source criticism, his scientific 'restraint' and his reluctance to be an educator, which made him the founder of 'objectivity' in Dutch historiography. It is interesting that among later generations of Dutch historians the perception of Fruin's method of research (*geschiedvorsing*) and his writing (*geschiedschrijving*) started to diverge. Whereas historians could no longer agree with his interpretation of the past and his (too narrow) identification of history with political history, they still regarded his scientific methodology as exemplary. This is why Dutch historians still consider him to be the 'father' of their discipline (Paul 2010; Tollebeek 1990, 15–76). 'We still see Fruin as a professional colleague (*vakgenoot*)', Herman Paul and Henk te Velde write in their introduction to a recent volume they edited on Fruin (Paul and Te Velde 2010, 14). This is certainly not the case with Van Heusde. The emergence of objectivity implied a crucial move towards academic historiography largely as we still know it today.

Fruin himself realized that scientific restraint had its limitations. In his inaugural lecture called The impartiality of the historian, he argued that the historian would never be able to create a 'mirror image' of history. A good historian therefore inevitably had to use his creativity and imagination to interpret his sources. Later he developed a more restrictive view with respect to allowing such 'subjective' elements. Fruin's own careful and detailed study of historical sources never led to a synthesis. What troubled historians was no longer the (hard-won) ability to make judgments, but how to deal with the principally problematic 'subjective' elements, which, as Fruin never lost sight of, were impossible to expel altogether (Tollebeek 1990, 49–56; Tollebeek 2010, 25).

Critics of Fruin's objectity feared that his methodological approach would remove all artistic elements from history. At his retirement Fruin was criticized by the young writer and poet Albert Verwey (1865–1937). Verwey belonged to the *Tachtigers*, a group of writers who cultivated literature as a purely esthetic and subjective expression of the artist (l'art pour l'art). 'Only the poet', Verwey believed, 'who does not spurn Intuition and Imagination (in capitals!), but applies it, can offer a true representation of past *Life* and evoke a large and simple image of the past. To think that the same can be attained by a profound study of the sources, is an illusion' (Tollebeek 1990, 54, paraphrasing Verwey). What this passage illustrates is that the rise of 'objective' science (and governance) went hand in hand with the emergence in the same period of the artist as a cultivator of the personal and subjective: side by side with the 'scientific self', the opposite 'artistic self' appeared (Daston and Galison 2007, 37).

## Disciplines and Professionalization

In the same way as the emergence of the statistical mindset occurred in a new socio-political setting, this also happened with scientific objectivity. The organization of learned culture changed dramatically. Learned societies decayed, and with them the local orientation of scholars and scientists. Increasingly, the universities monopolized scientific practice. A leading academic like Donders declared that societies had become obsolete: the road to scientific insight, as he stated, is '…not the way of societies. It is that of individuals'. A distance was created between the professional scientist and his (amateur) audience (Maas forthcoming; Weiss 2013). The scholars instead began to look for a community of peers, who were not confined to local boundaries, and organized themselves in scientific disciplines, with their own meetings, journals and practices.[9] As Frans van Lunteren has shown, whereas physics professors in the beginning of the century identified themselves as 'professor' (*hoogleraar*), they took to referring to themselves as physicists at the end of the century, indicating an increasing identification with their discipline (Van Lunteren 1995).

It should be noted that in the case of the humanities, disciplines never became as clearly delineated as was the case in the natural sciences. Disciplines in the humanities, in the words of Jo Tollebeek, remained 'flexible' and never completely became an academic monopoly. Often their fortunes were determined by current social and cultural developments (Tollebeek 2013).[10] Fruin's publications were also read by the general public, not only by professional colleagues (Aerts 2010). Nevertheless, although lacking a robust theoretical project as a disciplinary focal point, historians agreed upon a number of professional norms: sound historical work was grounded in the study of sources, historians should keep to the factuality of the events and exercise a certain degree of restraint when it came to making judgments and drawing conclusions (Tollebeek 2013, 83–85). In other words, the norms that made up the (loose) disciplinary identity were not very different from those that formed the essence of historical objectivity (which as has been mentioned is also less rigid than scientific objectivity).

David Cahan has noted that as in the nation state, the paradox of 'imagined communities' makes its presence felt in a scientific discipline: both consist of people who apparently do not have a lot in common, but still have a strong sense of belonging together (Cahan 2003, esp. 325–328). The language of 'objectivity' appears to be suitable, then, to bind together between people in 'imagined communities'.

The formation of scientific disciplines was part of a broader process of professionalization. From the middle of the nineteenth century, emerging professional groups like physicians and engineers started to found their own (national) organizations to promote their interests (Houwaart 1991; Lintsen 2005). Based on the specific

---

[9] On the formation of academic disciplines in nineteenth-century Netherlands: Tollebeek 2013; Van Lunteren 2013.

[10] For the historical discipline in the Netherlands, (see Aerts 2010).

skills they acquired during their training, professionals aspired to an autonomous position in society. The liberal reforms paved the way for this. Characteristically, a professional (ideally) owed his position to his specific skills and achievements, not to descent or wealth. The rise of the professional reflected the emergence of a socially more open and meritocratic society. As Pieter Caljé has argued for the case of the Netherlands, this development was also expressed in the Higher Education Act of 1876, which entailed that universities were no longer exclusively biased on 'elite reproduction' by general instruction but also focused on professional training of ambitious social climbers. Unsurprisingly, professional groups generally supported the liberal agenda. Fruin himself was a staunch supporter of Thorbecke.[11]

Another characteristic of professional groups is that they do not limit themselves to solving problems using only pre-existing knowledge. They employ their skills to deal with situations for which there is no ready-made solution at hand (Franzmann et al. 2015, 17–18). Thus the physician deals with illnesses the course of which is uncertain, and likewise academics such as Fruin dealt with empirical material that challenged the validity of larger, theoretical views. Wachelder has argued that the mindset for professionalization, namely a fundamental awareness of the open-endedness and unpredictability of the future, was, in his view, only realized among the Dutch in the second half of the nineteenth century, when the hierarchical social order that was regarded as basically unchangeable gradually transformed into a society in which one's place was no longer fixed (Wachelder 1992).

Fruin's 'objective' historiography, in short, was embedded in a society in which scientific disciplines, as well as professional groups, acquired a more or less autonomous position. Trust was vested to a lesser degree in the intuitive insights of 'harmonious', sage persons, and more in 'impersonal' practical skills to solve problems connected to a certain field of specialization.

## Conclusion

The Thorbeckian reforms signified a crucial moment in the modernization of Dutch society (i.e. in the development towards the Dutch society as we still know it today). The framework of this 'modern' society is that of an integrated nationstate, ruled in a democratic and formally bureaucratic manner. It is characterized by a relatively meritocratic and individualistic social structure. Within this society, modern scientific institutions likewise flourished: the roots of the research university and the research laboratory date back to the final decades of the nineteenth century, which is also true for the disciplines into which scientific practice is still currently organized. After Thorbecke's reforms – sometimes aptly called the 'art of separation' (of powers as well as of social domains) –several new social roles also emerged in

[11] For the professionalization and social rise of academic groups (see Caljé 2009, 340–358 and 458–512). Ironically, the 'liberal' Higher Education Act was implemented by the conservative minister Jan Heemskerk (1818–1897).

the cultural and scholarly domains which had been absent, or at least less clearly pronounced, before: thus with the creation of the professional researcher, the amateur appeared on the scene, and in the same period the artist and the 'intellectual' (as we still basically know them) also came to the fore (Aerts 1998). The scholar from the times of Van Heusde was, as it were, divided up in different personae.

In this period, too, objectivity emerged as a leading epistemic virtue. As I have sought to argue, its emergence both coincided with and was an expression of the societal reforms of the second half of the nineteenth century. In this era, however, as well as in later periods, its primacy was never unchallenged. The limits of objectivity have been subject to continuous debate, giving rise for instance to the 'neo-Kantian discussions' on the methodological distinction between the sciences and the humanities after 1875. In the view of Daston and Galison, objectivity later came to coexist with a new epistemic virtue: 'trained judgment'. However, as they also maintain, objectivity was there to stay. Indeed, both in science and in matters of governance of the bureaucratic nation-state, its techniques and protocols have persisted. Objectivity, and its attitude of scientific restraint, are inherently connected to 'modern' scientific practice, as well as 'modern' governance. In contrast to their 'sage' predecessors, pioneers of objectivity like Donders, Fruin, Kamerlingh Onnes and Kaiser are still recognized as pioneers, or even founding fathers of their respective disciplines.

Thorbecke spent many years as a misunderstood scholar. His scientific ideas, style and values were incompatible with those of the Dutch scholarly community in general and its figurehead Philip Willem van Heusde in particular. With a baffling consistency in actions, ideas and principles, the politician Thorbecke created the social-political circumstances in which the (objective) scholar Thorbecke would have flourished. The scholar to reap Thorbecke's fruits – as, so to speak, his scientific reincarnation – would be Robert Fruin. In him, and in many generations of historians that followed, the spirit of the scholar Thorbecke lives on. Van Heusde on the other hand has fallen into oblivion. His eclecticism, his lack of originality, his long-winded rhetoric, they all seem alien to us – a 'dreary street' in Rotterdam still bears his name (De Valk 1992: 13). This is Thorbecke's ultimate revenge.

## Bibliography

Aerts, R. 1997, *De letterheren: liberale cultuur in de negentiende eeuw: het tijdschrift* De Gids. Amsterdam: Meulenhoff.

———. 1998. Bevoegde autoriteiten: burgerlijke intellectuelen in de negentiende eeuw: een groepsportret. *De negentiende eeuw* 21: 72–95.

———. 2002. Burgerlijk sciëntisme? Over wetenschap en burgerlijke cultuur 1840–1880. In *Geleerden en leken: de wereld van de Hollandsche Maatschappij der Wetenschappen*, ed. R. Aerts, 17–31. Rotterdam: Werkgroep Negentiende Eeuw.

———. 2010. De lege kathedraal. In *Het vaderlandse verleden: Robert Fruin en de Nederlandse geschiedschrijving*, ed. H.J. Paul and H. te Velde, 195–220. Amsterdam: Bert Bakker.

[Anon]. 1840. Van Heusde, geschetst als geschiedkundige en als leeraar der geschiedenis: levensberigt van wijlen den hoogleeraar Philip Willem van Heusde door zijnen Ambtgenoot mr. Antonie van Goudoever. In *Utrechtsche studentenalmanak 1839*. Utrecht: Van Terveen.
Bagge, P. 1998. *Vorming door wetenschap: universitair onderwijs in Nederland 1815–1960*. Delft: Eburon.
Cahan, D. 2003. Looking At Nineteenth Century Science. In *From Natural Philosophy to the Sciences: Writing the History of Nineteenth Century Science*, ed. D. Cahan, 3–15. Chicago: University of Chicago Press.
Caljé, P. 2009. *Student, universiteit en samenleving: de Groningse universiteit in de negentiende eeuw*. Hilversum: Verloren.
Cossee, E.H. 1988. *Abraham des Amorie van der Hoeven 1798–1855*. Kampen: Kok.
Daston, L. 2014. Objectivity and impartiality: epistemic virtues in the humanities. In *The making of the humanities*, ed. T. Weststeijn, R. Bod, and J. Maat, vol. 3, 27–52. Amsterdam: Amsterdam University Press.
Daston, L., and P. Galison. 2007. *Objectivity*. New York: Zone Books.
De Haan, I., and H. te Velde. 1996. Vormen van politiek: veranderingen van de openbaarheid in Nederland 1848–1900. *Bijdragen en mededelingen betreffende de geschiedenis der Nederlanden* 111: 165–200.
De Quay, P.P. 2000. *De genoegzaamheid van het natuurlijk gezond verstand: prijsverhandelingen over godsdienst, zedenkunde en burgerlijke maatschappij in Nederland aan het einde der 18e eeuw*. The Hague: Sdu.
De Rooy, P. 2002. *Republiek van rivaliteiten: Nederland sinds 1813*. Amsterdam: Mets & Schilt.
De Valk, J.M.M. 1989. Philip Willem van Heusde. In *Philips Willem van Heusde: wijsbegeerte van het gezond verstand*, ed. J.M.M. de Valk, 13–35. Baarn: Ambo.
De Valk, J.P. 1992. Landsvader en landspaus? Achtergronden op de visie op kerk en school bij koning Willem I (1815–1830). In *Staats- en natievorming in Willem I's koninkrijk (1815–1830)*, ed. C.A. Tamse and E. Witte, 76–97. Brussels: Vubpress.
Drentje, J. 2004. *Thorbecke: een filosoof in de politiek*. Amsterdam: Boom.
Franzmann, A., A. Jansen, and P. Münte. 2015. Legitimizing science: Introductory essay. In *Legitimizing science: National and global publics (1800–2010)*, ed. A. Jansen, A. Franzmann, and P. Münte, 11–35. Chicago: University of Chicago Press.
Houwaart, E.S. 1991. *De hygiënisten: artsen, staat & volksgezondheid in Nederland 1840–1890*. Groningen: Historische Uitgeverij.
Huizinga, J. 1951. De geschiedenis der universiteit gedurende de derde eeuw van haar bestaan, 1814–1914. In *Verzamelde werken*, ed. J. Huizinga, vol. 8, 33–339. Haarlem: Tjeenk Willink.
Johannes, G.H. 1997. *De lof der aalbessen: over (Noord-)Nederlandse literatuurtheorie, literatuur en de consequenties van kleinschaligheid, 1770–1830*. The Hague: Sdu.
Klep, P.M.M. 2008. Governmentality, statistics and state power: Dutch labour and agricultural inquiries (1840–1914). In *The statistical mind in modern society: The Netherlands 1850–1940*, ed. P.M.M. Klep, J.G.S.J. Maarseveen, and I.H. Stamhuis, vol. 1, 261–287. Amsterdam: Aksant.
Klep, P.M.M, J.G.S.J. Maarseveen, and I.H. Stamhuis, eds. 2008. *The statistical mind in modern society: The Netherlands 1850–1940*, 2 vols. Amsterdam: Aksant.
Kloek, J., and W. Mijnhardt. 2001. *1800: blauwdrukken voor een samenleving*. The Hague: Sdu.
Knippenberg, H., and B. de Pater. 1990. *De eenwording van Nederland: schaalvergroting en integratie sinds 1800*. 2nd ed. Nijmegen: SUN.
Kossmann, E.H. 1995. *Vergankelijkheid en continuïteit: opstellen over geschiedenis*. Amsterdam: Bert Bakker.
———. 1987. Thorbecke en het historisme. In *Kossmann, Politieke theorie en geschiedenis: verspreide opstellen en voordrachten*, 306–347. Amsterdam: Bert Bakker.
Krop, P.C. 1992. *Legisme en privaatwetenschap: legisme in de Nederlandse privaatwetenschap in de negentiende eeuw*. 2nd ed. Deventer: Kluwer.
Lakke, A.J. 1908. *Philip Willem van Heusde*. Leiden: Brill.

Lintsen, H. 2005. *Ingenieurs in Nederland in de negentiende eeuw: een streven naar erkenning en macht*. Zutphen: Walburg Pers.
Lok, M. 2014. 'Tegen de partijgeest': bestuurlijke netwerken in het Verenigd Koninkrijk. In *Het (on)Verenigd Koninkrijk 1815–1850: een politiek experiment in de Lage Landen*, ed. R. Aerts and G. Deneckere, 61–68. The Hague: Algemeen Nederlands Verbond.
Maas, A. 2010. Civil scientists: Dutch scientists between 1750 and 1850. *History of Science* 48: 75–103.
———. Forthc. Sociëteit, moderniteit en wetenschap in eenzaamheid. *Archief van het Koninklijk Zeeuwsch Genootschap*.
Paul, H.J. 2010. De Hollandsche meester der streng-analytischen methode: Robert Fruin als vader van de Nederlandse geschiedwetenschap. In *Het vaderlandse verleden: Robert Fruin en de Nederlandse geschiedschrijving*, ed. H.J. Paul and H. te Velde, 221–248. Amsterdam: Bert Bakker.
———. 2011. Een Leids historisch ethos? De epistemische deugden van Fruin en Acquoy. *Leidschrift* 25: 95–114.
———. 2012. The Scholarly Self: Ideals of Intellectual Virtues in Nineteenth-Century Leiden. In *The Making of the Humanities*, ed. T. Weststeijn, R. Bod, and J. Maat, vol. 3, 397–411. Amsterdam: Amsterdam University Press.
Paul, H.J., and H. te Velde. 2010. Inleiding. In *Het vaderlandse verleden: Robert Fruin en de Nederlandse geschiedschrijving*, ed. H.J. Paul and H. te Velde, 7–15. Amsterdam: Bert Bakker.
Porter, Th. 1996. *Trust in numbers: The pursuit of objectivity in science and public life*. Princeton: Princeton University Press.
Roelevink, J. 1986. *Gedicteerd verleden: het onderwijs in de algemene geschiedenis aan de Universiteit te Utrecht, 1735–1839*. Amsterdam: APA-Holland Universiteits Pers.
Sassen, F. 1959. *Geschiedenis van de wijsbegeerte in Nederland tot het einde der negentiende eeuw*. Amsterdam: Elsevier.
Smit, J.W. 1958. *Fruin en de partijen tijdens de Republiek*. Groningen: Wolters.
Tamse, C.A., and E. Witte, eds. 1992. *Staats- en natievorming in Willem I's koninkrijk (1815–1830)*. Brussels: Vubpress.
Te Velde, H. 1992. *Gemeenschapszin en plichtbesef: liberalisme en nationalisme in Nederland, 1870–1918*. The Hague: Sdu.
———. 1999. Van grondwet tot grondwet, 1848–1917: oefenen met parlement, partij en schaalvergroting. In *Land van kleine gebaren: een politieke geschiedenis van Nederland 1780–1990*, ed. R. Aerts et al., 99–178. Nijmegen: SUN.
———. 2010. Politieke geschiedenis en synthetische kracht bij Robert Fruin. In *Het vaderlandse verleden: Robert Fruin en de Nederlandse geschiedschrijving*, ed. H.J. Paul and H. te Velde, 38–59. Amsterdam: Bert Bakker.
Theunissen, B. 2000. *'Nut en nog eens nut': wetenschapsbeelden van Nederlandse natuuronderzoekers, 1800–1900*. Hilversum: Verloren.
Tollebeek, J. 1990. *De toga van Fruin: denken over geschiedenis in Nederland sinds 1860*. Amsterdam: Wereldbibliotheek.
———. 2010. Een wetenschap van kleine gebaren: historiografische praktijken in de late negentiende eeuw. In *Het vaderlandse verleden: Robert Fruin en de Nederlandse geschiedschrijving*, ed. H.J. Paul and H. te Velde, 17–37. Amsterdam: Bert Bakker.
———. 2011. *Men of Character: The Emergence of the Modern Humanities*. Wassenaar: Netherlands Institute for Advanced Studies in the Humanities and Social Sciences.
———. 2013. Disciplines en studies: vernieuwing in de geesteswetenschappen. *Studium* 6: 79–90.
———. 2015. Fruin's aristocracy: historiographic practices in the late nineteenth century. In *The practice of philology in the nineteenth-century Netherlands*, ed. T. van Kalmthout and H. Zuidervaart, 79–101. Amsterdam: Amsterdam University Press.
Van Berkel, K. 2008. *De stem van de wetenschap: geschiedenis van de Koninklijke Akademie van Wetenschappen*. Vol. 1. Amsterdam: Bert Bakker.

Van Delft, D. 2005. *Heike Kamerlingh Onnes: een biografie: de man van het absolute nulpunt*. Amsterdam: Bert Bakker.

Van den Berg, W. 1984. Sociabiliteit, genootschappelijkheid en de orale cultus. In *Historische letterkunde. Facetten van vakbeoefening*, ed. M. Spies and W. van den Berg, 151–170. Groningen: Wolters Noordhoff.

Van Driel, J. 2016. The filthy and the fat: oeconomy, chemistry and resource management in the age of revolutions, 1700–1850. PhD thesis, University of Twente.

Van Heusde, P.W. 1860. *De Socratische school of wijsgeerte voor de negentiende eeuw*, vol. 1, 3rd ed. Groningen: H. Geertsema Jr. en J.B. Huber.

Van Holthoon, F.L. 1977. Mensen in Europa: ontwerp voor een sociale geschiedenis van Europa na 1750. Alphen aan den Rijn: Samsom Uitgeverij.

Van Lunteren, F. 2013. Het ontstaan van het systeem van bètadisciplines: de natuurkunde. *Studium* 6: 91–112.

———. 1995. Van meten tot weten': de opkomst der experimentele fysica aan de Nederlandse universiteiten in de negentiende eeuw. *Gewina* 18: 102–138.

Van Raak, R. 1999. Socratische wijsbegeerte voor het Nederlandse volk: Philip Willem van Heusde en het conservatisme in Nederland. In *Conservatisme in Nederland ca. 1780–1940*, ed. R. van Raak, 23–37. Leende: Damon.

Van Sas, N.C.F. 2005. *De metamorfose van Nederland: van oude orde naar moderniteit 1750–1900*. Amsterdam: Amsterdam University Press.

Vermeulen, E.E.G. 1959. *Smit over Fruin: kanttekeningen bij J.W. Smit, Fruin en de partijen tijdens de Republiek*. Utrecht: Dekker en Van de Vegt.

Vree, J. 1984. *De Groninger godgeleerden: de oorsprong en de eerste periode van hun optreden*. Kampen: Kok.

Wachelder, J. 1992. *Universiteit tussen vorming en opleiding: de modernisering van de Nederlandse universiteiten in de negentiende eeuw*. Hilversum: Verloren.

Weiss, M. P. M. 2013. *The Masses and the Muses: A History of Teylers Museum in the Nineteenth Century*. PhD thesis, Leiden University.

Zuidervaart, H., et al. 2011. *Frederik Kaiser (1808–1872): schepper van de 'nieuwe' Leidse Sterrewacht*. The Hague: Genootschap Gewina and Huygens ING.

# Index





J. van Dongen, H. Paul (eds.), *Epistemic Virtues in the Sciences and the Humanities*,
Boston Studies in the Philosophy and History of Science 321,
https://doi.org/10.1007/978-3-319-48893-6